Intelligent Control System

Applied Optimization

Volume 60

*Series Editors:*

Panos M. Pardalos
*University of Florida, U.S.A.*

Donald Hearn
*University of Florida, U.S.A.*

***The titles published in this series are listed at the end of this volume.***

# Intelligent Control Systems

## An Introduction with Examples

by

Katalin M. Hangos

*Department of Computer Science,*
*University of Veszprém,*
*Systems and Control Laboratory,*
*Computer and Automation Research Institute of the Hungarian Academy of Sciences*

Rozália Lakner

*Department of Computer Science,*
*University of Veszprém*

and

Miklós Gerzson

*Department of Automation,*
*University of Veszprém*

KLUWER ACADEMIC PUBLISHERS
DORDRECHT / BOSTON / LONDON

A C.I.P. Catalogue record for this book is available from the Library of Congress.

ISBN 1-4020-0134-7

---

Published by Kluwer Academic Publishers,
P.O. Box 17, 3300 AA Dordrecht, The Netherlands.

Sold and distributed in North, Central and South America
by Kluwer Academic Publishers,
101 Philip Drive, Norwell, MA 02061, U.S.A.

In all other countries, sold and distributed
by Kluwer Academic Publishers,
P.O. Box 322, 3300 AH Dordrecht, The Netherlands.

*Printed on acid-free paper*



Printed in the Netherlands.

**To our Better Halves**

# Contents


Acknowledgments xiii
Preface xv

1. GETTING STARTED 1
   1. Intelligent control: what does it mean? 2
   2. Components of intelligent control systems 3
      2.1 Software elements 3
      2.2 Users 5
   3. The structure and use of the book 6
      3.1 The structure of the material 6
      3.2 Prerequisites and potential readers 7
      3.3 Course variants 8

2. KNOWLEDGE REPRESENTATION 11
   1. Data and knowledge 12
      1.1 Data representation and data items in traditional databases 12
      1.2 Data representation and data items in relational databases 14
   2. Rules 15
      2.1 Logical operations 15
      2.2 Syntax and semantics of rules 18
      2.3 Datalog rule sets 19
         2.3.1 The dependence graph of datalog rule sets 21
   3. Objects 22
   4. Frames 26
   5. Semantic nets 27

3. REASONING AND SEARCH IN RULE-BASED SYSTEMS 31
   1. Solving problems by reasoning 31
      1.1 The structure of the knowledge base 32
      1.2 The reasoning algorithm 33
      1.3 Conflict resolution 36

1.4 Explanation of the reasoning 38
2. Forward reasoning 38
2.1 The method of forward reasoning 38
2.2 A simple case study of forward reasoning 41
3. Backward reasoning 44
3.1 Solving problems by reduction 44
3.2 The method of backward reasoning 45
3.3 A simple case study of backward reasoning 48
4. Bidirectional reasoning 51
5. Search methods 51
5.1 The general search algorithm 52
5.2 Depth-first search 53
5.3 Breadth-first search 54
5.4 Hill climbing search 55
5.5 A* search 56

4. VERIFICATION AND VALIDATION OF RULE-BASES 59
1. Contradiction freeness 60
1.1 The notion of contradiction freeness 60
1.2 Testing contradiction freeness 61
1.3 The search problem of contradiction freeness 63
2. Completeness 64
2.1 The notion of completeness 64
2.2 Testing completeness 64
2.3 The search problem of completeness 65
3. Further problems 66
3.1 Joint contradiction freeness and completeness 66
3.2 Contradiction freeness and completeness in other types of knowledge bases 66
4. Decomposition of knowledge bases 67
4.1 Strict decomposition 68
4.2 Heuristic decomposition 68

5. TOOLS FOR REPRESENTATION AND REASONING 69
1. The Lisp programming language 70
1.1 The fundamental data types in Lisp 70
1.2 Expressions and their evaluation 72
1.3 Some useful Lisp primitives 73
1.3.1 The QUOTE primitive 73
1.3.2 Primitives manipulate on lists 74
1.3.3 Assignment primitives 76
1.3.4 Arithmetic primitives 76
1.3.5 Predicates 77
1.3.6 Conditional primitives 79
1.3.7 Procedure definition 81
1.4 Some simple examples in Lisp 82
1.4.1 Logical functions 82
1.4.2 Calculating sums 83

1.4.3 Polynomial value 84
2. The Prolog programming language 84
2.1 The elements of Prolog programs 85
2.1.1 Facts 85
2.1.2 Rules 87
2.1.3 Questions 87
2.1.4 The Prolog program 88
2.1.5 The declarative and procedural views of a Prolog program 89
2.1.6 More about lists 89
2.2 The execution of Prolog programs 90
2.2.1 How questions work 90
2.2.2 Unification 92
2.2.3 Backtracking 93
2.2.4 Tracing Prolog execution 94
2.2.5 The search strategy 95
2.2.6 Recursion 96
2.3 Built-in predicates 96
2.3.1 Input-output predicates 97
2.3.2 Dynamic database handling predicates 97
2.3.3 Arithmetic predicates 98
2.3.4 Expression-handling predicates 98
2.3.5 Control predicates 99
2.4 Some simple examples in Prolog 99
2.4.1 Logical functions 99
2.4.2 Calculation of sums 100
2.4.3 Path finding in a graph 101
3. Expert system shells 103
3.1 Components of an expert system shell 104
3.2 Basic functions and services in an expert system shell 105

6. REAL-TIME EXPERT SYSTEMS 109
1. The architecture of real-time expert systems 110
1.1 The real-time subsystem 111
1.2 The intelligent subsystem 113
2. Synchronization and communication between real-time and intelligent subsystems 114
2.1 Synchronization and communication primitives 114
2.2 Priority handling and time-out 115
3. Data exchange between the real-time and the intelligent subsystems 116
3.1 Loose data exchange 117
3.2 The blackboard architecture 119
4. Software engineering of real-time expert systems 121
4.1 The software lifecycle of real-time expert systems 122
4.2 Special steps and tools 125

7. QUALITATIVE REASONING 127

1. Sign and interval calculus 128
1.1 Sign algebra 129
1.2 Interval algebras 130
2. Qualitative simulation 132
2.1 Constraint type qualitative differential equations 132
2.2 The solution of QDEs: the qualitative simulation algorithm 138
2.2.1 Initial data for the simulation 138
2.2.2 Steps of the simulation algorithm 139
2.2.3 Simulation results 142
3. Qualitative physics 145
3.1 Confluences 145
3.2 The use of confluences 147
4. Signed directed graph (SDG) models 148
4.1 The structure graph of state-space models 148
4.2 The use of SDG models 151

8. PETRI NETS 153
1. The Notion of Petri nets 154
1.1 The basic components of Petri nets 154
1.1.1 Introductory examples 154
1.1.2 The formal definition of Petri nets 162
1.2 The firing of transitions 162
1.3 Special cases and extensions 165
1.3.1 Source and sink transitions 165
1.3.2 Self-loop 165
1.3.3 Capacity of places 166
1.3.4 Parallelism 168
1.3.5 Inhibitor arcs 172
1.3.6 Decomposition of Petri nets 175
1.3.7 Time in Petri nets 176
1.4 The state-space of Petri nets 177
1.5 The use of Petri nets for intelligent control 178
2. The analysis of Petri nets 178
2.1 Analysis Problems for Petri Nets 179
2.1.1 Safeness and Boundedness 179
2.1.2 Conservation 179
2.1.3 Liveness 180
2.1.4 Reachability and Coverability 180
2.1.5 Structural properties 180
2.2 Analysis techniques 181
2.2.1 The reachability tree 181
2.2.2 Analysis with matrix equations 186

9. FUZZY CONTROL SYSTEMS 191
1. Introduction 191
1.1 The notion of fuzziness 191
1.2 Fuzzy controllers 192
2. Fuzzy sets 192

2.1 Definition of fuzzy sets 192
2.2 Operations on fuzzy sets 200
2.2.1 Primitive fuzzy set operations 201
2.2.2 Linguistic modifiers 205
2.3 Inference on fuzzy sets 208
2.3.1 Relation between fuzzy sets 209
2.3.2 Implication between fuzzy sets 211
2.3.3 Inference on fuzzy sets 214
3. Rule-based fuzzy controllers 215
3.1 Design of fuzzy controllers 216
3.1.1 The input and output signals 216
3.1.2 The selection of universes and membership functions 217
3.1.3 The rule-base 219
3.1.4 The rule-base analysis 220
3.2 The operation of fuzzy controllers 223
3.2.1 The preproccessing unit 223
3.2.2 The inference engine 223
3.2.3 The postprocessing unit 225

10. G2: AN EXAMPLE OF A REAL-TIME EXPERT SYSTEM 227
1. Knowledge representation in G2 228
2. The organization of the knowledge base 230
2.1 Objects and object definitions 231
2.2 Workspaces 232
2.3 Variables and parameters 233
2.4 Connections and relations 234
2.5 Rules 235
2.6 Procedures 237
2.7 Functions 238
3. Reasoning and simulation in G2 239
3.1 The real-time inference engine 239
3.2 The G2 simulator 240
4. Tools for developing and debugging knowledge bases 241
4.1 The developers' interface 241
4.1.1 The graphic representation 241
4.1.2 G2 grammar 242
4.1.3 The interactive text editor 242
4.1.4 The interactive icon editor 243
4.1.5 Knowledge base handling tools 244
4.1.6 Documenting in the knowledge base 245
4.1.7 Tracing and debugging facilities 246
4.1.8 The access control facility 247
4.2 The end-user interface 247
4.2.1 Displays 247
4.2.2 End-user controls 248
4.2.3 Messages, message board and logbook 249
4.3 External interface 250

Appendices 251

A– A BRIEF OVERVIEW OF COMPUTER CONTROLLED SYSTEMS 251

1. Basic notions in systems and control theory 251
   - 1.1 Signals and signal spaces 252
   - 1.2 Systems 252
2. State-space models of linear and nonlinear systems 253
   - 2.1 State-space models of LTI systems 254
   - 2.2 State-space models of nonlinear systems 254
   - 2.3 Controllability 255
   - 2.4 Observability 256
   - 2.5 Stability 257
3. Common functions of a computer controlled system 258
   - 3.1 Primary data processing 258
   - 3.2 Process monitoring functions 260
   - 3.3 Process control functions 260
   - 3.4 Functional design requirements 262
4. Real-time software systems 262
   - 4.1 Characteristics of real-time software systems 262
   - 4.2 Elements of real-time software systems 264
   - 4.3 Tasks in a real-time system 264
5. Software elements of computer controlled systems 268
   - 5.1 Characteristic data structures of computer controlled systems 268
     - 5.1.1 Raw measured data and measured data files 269
     - 5.1.2 Primary processing data file 270
     - 5.1.3 Events data file 270
     - 5.1.4 Actuator data file 271
   - 5.2 Typical tasks of computer controlled systems 272
     - 5.2.1 Measurement device handling 272
     - 5.2.2 Primary and secondary processing 272
     - 5.2.3 Event handling 272
     - 5.2.4 Controller(s) and actuator handling 273

B– THE COFFEE MACHINE 275

1. System description 275
2. Dynamic model equations 277
   - 2.1 Differential (balance) equations 278
   - 2.2 System variables 279

References 281

Index 289

About the Authors 301

# Acknowledgments

With the high popularity and expectations of intelligent control systems in our minds, we felt a great challenge to come up with a textbook in intelligent control systems. That is why we are particularly grateful for all those who have encouraged us to get through: our colleagues, students and families.

The material is based on our intelligent control course for 4th and 5th year information engineers in the University of Veszprém (Hungary) which has been taught successfully for 5 years for more than 100 students. The support of the University, our colleagues and students is gratefully acknowledged.

The inspiring and friendly atmosphere at the Department of Computer Science at the University of Veszprém and that of the Systems and Control Laboratory of the Computer and Automation Research Institute has also contributed to the writing of this book.

Special thanks to Gábor Szederkényi who helped us with all technical and LATEX problems.

# Preface

Disciplines are diverging and converging. That is a natural process of science. Diverging is the deeply penetrating characteristic of science, opening knowledge about new phenomena and creating new methods. Convergence emerges by the interaction of disciplines, it serves as a relevant driving force towards new more effective syntheses. Convergence is evoked by the subject itself, i.e. by science-supported solving of practical tasks.

Control of industrial processes is the best example. Physics, chemistry and mechanics join the control of dynamically changing processes and control methods as a result of mathematical system theory. We can enumerate several further relations, economy and sociology, the whole world of the process and the applying human being.

Here stops the university educator in writing a textbook: What are the constituents of the basic knowledge for an engineer to be prepared for intelligent control? What are easily digestible, stemming from earlier courses? Where should his/her own course be ended, hoping that the further studies and especially the diligence and practice of the student enhances all these for enabling to complete the realistic, highly complex tasks of intelligent process modeling, design and control? That means the thorough and, on the other hand, general knowledge of system requirements.

The underlying textbook is the result of several years teaching experience and could not be based on similar course books in the field. The reason is evident: dynamic system analysis and synthesis applied ideas of artificial intelligence in the past few years only. These methods relate to the general methods of representation functional dynamics, e.g. Petri-nets; different methods of handling uncertainty, especially in cases where statistics is not sufficient but human experience has a relevant role, e.g. fuzzy concept. The description of dynamics is more meaningful by qualitative methods due to discrete changes in the status and consistence of the materials concerned. Basic is the application of rules and logical reasoning in the analysis of phenomena and control operation. Special tools, such as programming languages dedicated for logical reasoning, shells for creating consultation systems in a special field, i.e. expert systems should be added, too.

The convergence of disciplines open a very suitable pedagogical means for examples related to the real life phenomena of those procedures where the student is familiar. By this way the reader receives much better insight into the subject, can understand theoretical concepts by his/her own personal impression that enables the stimulation of further steps outlined a little bit above.

I wish success for the textbook and to the students, started with this initiative!

Tibor Vámos
Member of the Hungarian Academy of Sciences
Computer and Automation Research Institute
Budapest, 21th June, 2001

Chapter 1

# GETTING STARTED

Intelligent control is a rapidly developing, complex and challenging field with great practical importance and potential. It emerged as an interdisciplinary field of computer controlled systems and artificial intelligence (AI) in the late seventies or early eighties when the necessary technical and theoretical infrastructure in both computer science and real-time computation techniques became available.

A great deal of interest has been shown in learning more about intelligent control by a wide audience. It has been a challenging and popular course subject for both graduate and undergraduate students of various engineering disciplines. At the same time there is a growing need amongst industrial practitioners to have textbook material on the subject readily to hand.

Because of the rapidly developing and interdisciplinary nature of the subject, the information available is mainly found in research papers, intelligent control system manuals and – last but not least – in the minds of practitioners, of engineers and technicians in various fields. There are a few edited volumes consisting of research papers on intelligent control systems [1], [2]. Little is known and published about the *fundamentals and the general know–how in designing, implementing and operating intelligent control systems*. Therefore, the subject is suitable mainly for elective courses on an advanced level where both the material and the presentation could and should be flexible: a core basic material is supplemented with variable parts dealing with the special tools and techniques depending on the interest and background of the participants.

# 1. INTELLIGENT CONTROL: WHAT DOES IT MEAN?

The notion of intelligent control systems is based on a joint understanding of the notions of "control systems" and "intelligent systems". Both of the above notions have undergone a strong development and have been the subject of disputes and discussions (see e.g. [3]). Therefore we shall restrict ourselves to practical, engineering type definitions of both, in describing the subject matter of this book.

*Control systems* assume the existence of a dynamic system to be controlled, that is an object the behaviour of which is time-dependent behaviour and which responds to the influences of its environment described by the so called input signals by output signals. The control system then senses both input and output and designs an input that achieves a predefined control aim.

Control systems are most often realized using computers, and in these cases we talk about *computer-controlled systems.* A computer-controlled system is by nature a real-time software system. Its software architecture contains standard data structures and tasks operating thereon. These include the following:

- *data structures:* raw measured data, measured data, events, etc.
- *tasks:* measurement device handling, primary processing, event handling, etc.

Appendix A gives a detailed description of the most important terms and notions in systems and control theory, as well as the software structure of a computer controlled system.

The notion of *intelligence* in the sense of *artificial intelligence* [4]-[8] is the other ingredient in the term "intelligent control systems". The notion of *intelligence* in itself has been a subject of permanent discussion for a long time and *artificial intelligence* is understood as "computer-aided intelligence", that is intelligence produced by computers.

The engineering type definition of artificial intelligence can be best understood if one recalls the elements of a problem for which we think we need a clever or "intelligent" solution. It is intuitively clear that easy or trivial tasks do not need a clever solution, just – perhaps – hard work. On the other hand, clever or intelligent solutions exhibit at least some non-trivial, surprising or unusual element, approach or other ingredient [9]. Therefore, one may say that an *intelligent method solves*

- *a difficult (non-trivial, complex, unusually large or complicated) problem*

- *in a non-trivial, human-like way.*

Furthermore, we can identify another basic characteristic of intelligent methods if we follow the idea of the engineering type definition above. The basic difference between the human and the machine way of solving difficult problems is that humans prefer to use clever heuristics over mechanistic exhaustive "brute force" approaches. The presence of *heuristics* is one of the key characteristics of intelligent methods.

To summarize we can say that *intelligent control systems are computer-controlled systems where at least part of the control tasks performed require intelligent methods.*

# 2. COMPONENTS OF INTELLIGENT CONTROL SYSTEMS

Every object with some kind of intelligence exhibits a quite complex and sophisticated structure: think of the biological structure of our nervous system controlled by our brain. Similarly, intelligent control systems have special components which are necessary to carry out control in an intelligent way. Most of the software elements of an intelligent control system perform its control function but some special elements serve its users, who come from various backgrounds and have varying academic qualifications.

## 2.1 SOFTWARE ELEMENTS

As we have already seen before, intelligent control systems are computer controlled systems with intelligent element(s) [10]. This implies that Neuman's principle applies to these systems: they have separate elements for the inherently passive, *data type* part and the active, *program type* part.

In traditional software systems, like in computer controlled systems, the data type elements are usually organized in a *database* while the active elements are *real-time tasks.* Tasks share the data in the database and a special task, the *database manager* is responsible for the resource management and the consistency of the data base.
This separation is clearly visible on the software structure of a computer controlled system described in details in section 5. in Appendix A.

Clearly not every intelligent system obeys Neuman's principle. Our brain, for example, works in a distributed manner, where every neuron has processing functions and stores data as well by connecting to other neurons.

The intelligent software systems that obey Neuman's principle are called *knowledge-based* systems. In intelligent software systems one can also find elements of the data and program type, therefore they are all knowledge-based systems. These elements, however, are given other special names as compared to traditional software systems.

The basic elements of a knowledge-based system are depicted in Fig. 1.1.

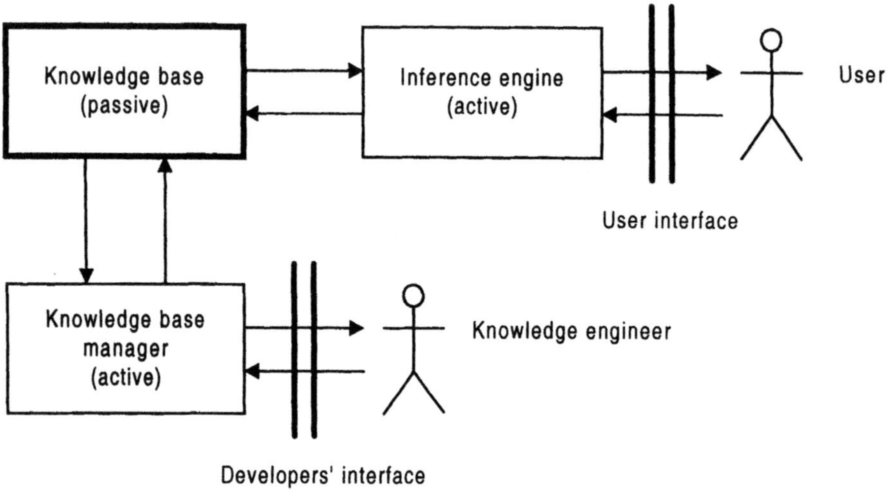


*Figure 1.1.* The structure of knowledge-based systems

We can see the following active and passive elements:

1. *Knowledge base*
   The database of a knowledge-based system is called the knowledge base. There is, however, a substantial difference between a database with data entirely passive and a knowledge base where the relationships between the individual data elements are much more important. We shall learn more about the similarities and differences between data and knowledge bases in Chapter 2.

2. *Inference engine*
   The inference engine of a knowledge-based system is its processing (program) element. It uses the content of the knowledge base to derive new knowledge items using the process of *reasoning*. Reasoning in rule-based expert systems is the subject of a separate chapter, Chapter 3.

There can be more than one inference engine in a knowledge-based system, in the same way as there are multitasking traditional software systems.

3. *Knowledge base manager*
   Similarly to the database manager, the knowledge base manager of a knowledge-based system performs the resource and consistency management of the knowledge base. However, this task is much more difficult than that of the database manager's, because the relationships between knowledge items are much more complex. As it is shown in Chapter 4 even checking the completeness and contradiction freeness of a rule-based knowledge base is computationally hard.

There is a special, important and widely used special type of knowledge-based systems where the knowledge is collected from an expert in a specific application domain. Such a knowledge-based system in a specific domain is called an *expert system.* If , in addition, the knowledge base contains data items and logical relationships between them expressed in the form of *rules* we speak about a *rule-based expert system* [11].

## 2.2 USERS

There are two principally different types of users in any knowledge-based system and their roles, qualification and user privileges are different.

1. *Knowledge engineer*
   A knowledge engineer is a person with a degree in computing, software engineering, programming or alike with specialization in intelligent systems. The design, implementation, verification and validation of a knowledge-based system is done by knowledge engineers. Ideally, they should have an interdisciplinary background knowing both knowledge-based systems technology and the application field in which the knowledge-based system is being used. In the case of intelligent control systems, a knowledge engineer should be familiar with the basic notions and principles of computer controlled systems as well.

   Knowledge engineers use the so called *developers´ interface* which is designed to work directly with the knowledge base manager of the knowledge-based system. Through this interface high privilege tasks, such as changing the structure and content of the knowledge base and other knowledge base management tasks can be carried out.

2. *User*
   A knowledge-based system is most often used via the so called *user*

*interface* which connects users to the inference engine. Users can ask questions to be answered and can initiate tasks to be performed with the use of reasoning. Various advanced user support functions, such as debugging, explanation, intelligent "what if" type hypothesis testing etc. are usually also offered by the user interface.

In order to protect the knowledge-based system from damages, malfunctions and inconsistency ordinary users have much fewer privileges than knowledge engineers. Therefore, there is usually no possibility for a user to change the structure of the knowledge base or to enter new knowledge item without a consistency check.

The role and place of these users in an intelligent control system can be seen in Fig. 1.1. *The general aim of this book is to provide the reader with the necessary knowledge and expertise to become a knowledge engineer of intelligent control systems.*

## 3. THE STRUCTURE AND USE OF THE BOOK

Keeping in mind that intelligent control is a rapidly developing area, we designed the structure of the book to be as flexible and modular as possible. This arrangement of the material makes it possible to use the book in various ways depending on the needs and background of the reader(s). Furthermore, it offers a possibility to combine the material presented here with other information about various tools and techniques not present in this book on intelligent control.

### 3.1 THE STRUCTURE OF THE MATERIAL

The textbook deals with the basic concepts and the most widely used tools and techniques in intelligent control illustrated by simple examples. Furthermore, it contains chapters dealing with some of the advanced tools and techniques applied in intelligent control systems. However, the authors' expertise, background and interest determined the selection, therefore some of the widely used techniques may be left out.

Most of the chapters contain tutorial material as well, either in separate sections and sub-sections or in the form of in-text illustrative examples. A large part of the tutorial examples is computer-based and uses the appropriate knowledge representation and reasoning tool. Some of them in Chapter 10 uses G2 of Gensym.

A simple process system example, a coffee machine, is used extensively in the book to illustrate the various tools and techniques. The system description and the development of the dynamic state space model of the coffee machine is found in Appendix B.

The material in the book is divided into three parts:

- *"core" background material* (Chapters 2-3)
  These chapters include basic information on knowledge representation and reasoning summarizing the relevant notions in intelligent control, together with the tools and techniques from the field of artificial intelligence. Familiarity with these in at least the depth presented here is necessary for any course in intelligent control systems.

- *advanced methods and tools for design, implementation and analysis* (Chapters 4-6)
  The problems and solution techniques in knowledge base validation and verification and the most common tools for knowledge representation and reasoning - including Lisp, Prolog and expert system shells, as well as the basic properties of real-time expert systems - are presented here. This part of the book is mainly dedicated to the future knowledge engineers and requires higher academic qualifications and background. Therefore some parts may be omitted or substantially shortened according to the readers' interest. At the same time, part of the material presented in these chapters belongs to the "core" knowledge in intelligent control systems.

- *special tools and techniques in intelligent control* (Chapters 7-10)
  Separate chapters are devoted to the following tools and techniques in intelligent control:

  – qualitative modelling
  – Petri nets
  – fuzzy control systems
  – G2: a real-time expert system of Gensym

  These chapters are largely independent of each other but depend on the previous chapters. As a consequence, these chapters can be read in any order and any of them can be omitted if necessary.

## 3.2 PREREQUISITES AND POTENTIAL READERS

The interdisciplinary and rapidly developing nature of the topic as well as the broad and diverse background of potential readers requires the prerequisites to be restricted to a necessary minimum. Only higher mathematics basics that are commonly taught at engineering faculties, such as linear algebra, elementary calculus, fundamentals of mathematical logics and combinatorics (graphs) are requested. Elementary notions in computers and computations such as data structures, algorithms and software engineering are advisable.

There are, however, two disciplines on which intelligent control heavily depends: artificial intelligence and computer controlled systems. The necessary background in artificial intelligence is summarized in Chapters 2-3. A brief overview of computer controlled systems is given in Appendix A.

## 3.3 COURSE VARIANTS

In approximately 300 pages *INTELLIGENT CONTROL SYSTEMS: An Introduction with Examples* aims to be a textbook for higher years undergraduate and graduate engineering students. It can not only be used by students attending elective courses but - for purposes of self-study - also by engineers who are already working and are interested in the subject.

The modular and flexible arrangement of the material in this book means that it can be used in different courses depending on the background and interest of the participants. The possible examples of how the material might be used are as follows.

1. *Introduction to Intelligent Control Systems*
   (an introductory course for higher level undergraduate engineering students)
   This course can be an elective course in intelligent control for final year engineering students presenting only the basic ideas. The aim of the course is to prepare them to be "educated" users of intelligent control systems and to help knowledge engineers to design, implement and operate intelligent control systems. The material of such a course may include

   - *"core" background material* (Chapters 2-3)
   - *a brief overview of computer controlled systems* (Appendix A)
   - a selection of the material from *advanced design, implementation and analysis methods and tools* (Chapters 4-6)
   - *G2* as an illustrative example (Chapter 10)

2. *Intelligent Control Systems*
   (graduate or post-graduate course for future knowledge engineers)
   The material of the book is primarily designed to be an "ideal" textbook for such a course, both in its content and the depth of presenting the material. However, if the lecturer has other preferences or experience related to the *special tools and techniques in intelligent control*

part, any of the chapters here may be omitted, extended or substituted by something else.

In particular, neural networks, which are highly popular in the field of intelligent control, have been omitted from the present version of the book. They can be covered by a graduate course at the price of leaving out qualitative modelling, for example.

3. *Fuzzy Techniques in Intelligent Control*
   (graduate or post-graduate course for engineers)
   The material presented in this book can serve as "core" material in any advanced intelligent control course focusing on a particular technique (fuzzy control, for example). In this case, the course contents may be the following.

   - a brief overview of the *background "core" material* (Chapters 2-3)
   - *a brief overview of computer controlled systems* (Appendix A)
   - *advanced methods and tools for design, implementation and analysis* (Chapters 4-6)
   - the relevant chapter amended by additional material on the particular technique in intelligent control (Chapter 9 and additional material in case of fuzzy techniques).

Chapter 2

# KNOWLEDGE REPRESENTATION

Knowledge bases are basic building elements of intelligent control systems. Therefore the understanding of the principles, methods and tools of knowledge representation is of vital importance. Knowledge items describe

1. data needed for the problem solving
2. relationships among data elements in the real world.

This chapter deals with knowledge representation methods [12], [13] as natural extensions to the traditional data representation methods [14].

Because of their theoretical and practical importance, special emphasis is put to rule-based systems (where rules are the main knowledge representation tools).

Knowledge representation methods, which are used for the organization, verification and validation of knowledge bases are also discussed in this chapter.

The material is arranged in the following sections.

- *Data and knowledge*
  The similarities and differences between data and knowledge and their representation methods.

- *Rules*
  Rules are the most common and most widely used knowledge representation tools. This section describes their syntax together with the properties of special rule-bases.

- *Objects*
  Objects are mainly used, when for structuring knowledge based sys-

tems therefore the main emphasis is put on their encapsulating properties here.

- *Frames*
  Frames can be seen as extensions of records with standard active elements. This view explains why they can effectively be used for knowledge representation.

- *Semantic nets*
  Semantic nets are graphic tools for describing semantic relationships between knowledge items in knowledge bases. The description highlights their use for knowledge base verification purposes.

# 1. DATA AND KNOWLEDGE

As we have already seen in section 2. of Chapter 1. the passive (executable) part of a knowledge based intelligent software system is stored in its knowledge base. This fact explains the similar role databases and knowledge bases play in software systems. The differences between data and knowledge and their representation methods originate from the higher complexity of knowledge as compared to data in a database.

In this section we briefly review the most important properties of data representation in traditional and more advanced relational databases in order to show how advanced data representation approaches may lead us to knowledge representation techniques.

In order to solve complex problems in an intelligent system we need a lot of information - data and knowledge - about the objects and their relationships in the real world and there is also a need for methods and algorithms that use this information for finding solutions to problems. The properties of the objects in the real world are described by *facts* or data and the connections or dependencies between these facts are given by *relationships*. In the following we will show how facts and relationships are described in traditional and relational databases.

## 1.1 DATA REPRESENTATION AND DATA ITEMS IN TRADITIONAL DATABASES

In a traditional database the set of related data items is stored in a *record*. The structure of a record type is fixed and it is defined in the declaration part of the program which uses these type of records. Records contain *fields* of fixed type for the data items in them.

A simple example of record declaration is given below. The record shown stores the data items belonging to raw measured data in a computer controlled system as explained in section 5. in Appendix A.

**EXAMPLE 2.1** A simple record type

Consider a simple record for storing the related data items of raw measured data in a computer controlled system declared in Pidgin Algol syntax.

```
raw-measurement  record
                 identifier:  string;
                 type:        character; {'R','B'}
                 value:       word;      {unscaled, type-dependent}
                 meas-time:   integer array[6];
                                         {ss-mm-hh-dd-mm-yy}
                 error-code:  word;      {type-dependent}
                 end;                    {raw-measurement}
```

A *file* is an ordered set of records of the same type. *The attributes of files in a traditional database are:*

- *identifier*
- *record type (structure)*
- *mode of use: read only, read/write etc.*
- *ordering: sequential, indexed etc.*
- *length: fixed (with maximal number of records), variable etc.*

A database is then the set of files.

In conclusion we can say that traditional databases are characterized by the following properties from the viewpoint of possible knowledge representation.

- Facts are stored in record fields that have a fixed structure.
- The possibilities to describe relationships are rather limited, this is done by the declaration of field types and by specifying default values.
- The data structures are completely passive, it is not possible to describe actions to be performed on the individual data items.

## 1.2 DATA REPRESENTATION AND DATA ITEMS IN RELATIONAL DATABASES

To overcome some of the limitations of traditional databases explained above, relational databases have been developed.

The properties of a relational database are as follows.

1. A set of related data items is stored in a *record* but here the record only defines the logical grouping of data items, which physically may be stored elsewhere. A record contains fields of fixed type and structure.

2. Default values and relationships can be specified as so called *relations* to any of the fields or to a group of fields. The relations can be of logical and/or arithmetic type. Relations can be defined for

   - the default and admissible values of a field,
   - the values of fields in the same record,
   - the values of fields in different records or different record types.

A simple example illustrates the properties above.

---

**EXAMPLE 2.2** A simple "active" record with a relation

Consider a simple record for storing the operands and result of an addition

$$a + b = c \tag{2.1}$$

```
add-rec record
        a:      real;    { op-1 }
        b:      real;    { op-2 }
        c:      real;    { result }
        end;    { add }
```

equipped with the relation (2.1).

The record will be accepted by the database manager if the relation holds. If one of the fields is missing, i.e. has the value **nil** then the database manager fills it in to satisfy the relation (2.1).

---

The example above shows that the *relations may call for an action which is performed automatically by the database manager if need arises.*

A set of relational records of the same structure forms a relational file. A relational database is then a set of relational files and the set of relations connecting them.

From the viewpoint of knowledge representation a relational database exhibits the following properties.

- It has a much more flexible structure than a conventional database.
- The database manager ensures the consistency of the database and the fulfillment of the relations, furthermore it provides the default values.
- Facts are stored in relational database records.
- Relationships are described using the relations.

The properties above explain why *knowledge bases can in principle be realized using relational databases.*

## 2. RULES

Rules are the most widespread form of knowledge representation in expert systems and other AI tools. Their popularity is explained by their simplicity and transparency from both a theoretical and a practical point of view. This implies that rule sets are relatively easy to handle and investigate.

As we shall see later in Chapter 4, the logical validation of a rule set, i.e. the check of its consistency and contradiction freeness is a hard problem from algorithmic viewpoint (the problem is not polynomial but NP-hard). Rule sets mostly describe black box type heuristic knowledge, therefore they are difficult to validate against other type of engineering knowledge, say against process models. There are some methods, however, based on qualitative process models for partial validation of this type as it is described later in section 3. of Chapter 7.

This section contains a short summary of logical operations in order to prepare the ground for describing the syntax and semantics of rules as well as to introduce a special type of rule sets.

### 2.1 LOGICAL OPERATIONS

The properties of the well–known logical operations are briefly summarized here in order to serve as a basis for defining the syntax of rules.

This subsection will also enable us to extend these operations towards the sign operations.

Logical variables in traditional logics may have two distinct logical constant values: **true** and **false**. The logical operations on these logical variables are defined by so called *operation tables*. The operation tables of logical operations are also called *truth tables*.

For example, the following truth tables in Table 2.1 and 2.2 define the logical **and** ($\wedge$) and **implication** ($\rightarrow$) operations.

*Table 2.1.* Operation table of the "and" operation

| $a \wedge b$<br>$a \downarrow \quad b \rightarrow$ | **false** | **true** |
|---|---|---|
| **false** | **false** | **false** |
| **true** | **false** | **true** |

*Table 2.2.* Operation table of the "implication" operation

| $a \rightarrow b$<br>$a \downarrow \quad b \rightarrow$ | **false** | **true** |
|---|---|---|
| **false** | **true** | **true** |
| **true** | **false** | **true** |

The logical operations ($\wedge$, $\vee$, $\neg$, $\rightarrow$) have the following well-known algebraic properties:

1. *commutativity*:

$$(a \wedge b) = (b \wedge a) \quad , \quad (a \vee b) = (b \vee a)$$

2. *associativity*:

$$(a \wedge b) \wedge c = a \wedge (b \wedge c) \quad , \quad (a \vee b) \vee c = a \vee (b \vee c)$$

3. *distributivity*:

$$a \wedge (b \vee c) = (a \wedge b) \vee (a \wedge c) \quad , \quad a \vee (b \wedge c) = (a \vee b) \wedge (a \vee c)$$

4. *de Morgan identities*:

$$\neg(a \wedge b) = \neg a \vee \neg b \quad , \quad \neg(a \vee b) = \neg a \wedge \neg b$$

With the logical identities above, every logical expression can be transformed into *canonical form*. There are three types of canonical forms:

- the *disjunctive normal form* or *DNF* is disjunction of conjunctions of atomic formulas (logical constants or logical variables or predicates) or their negations

  ex.

$$(\neg a \land b) \lor (c \land \neg d)$$

- the *conjunctive normal form* or *CNF* is conjunction of disjunctions of atomic formulas or their negations

  ex.

$$(\neg a \lor b) \land (c \lor \neg d)$$

- the *implicative normal form* or *INF* is an implication with the conjunction of atomic formulas on the left and disjunctions of atoms on the right

  ex.

$$(\neg a \land b) \rightarrow (c \lor \neg d)$$

Traditional two-valued logic is usually extended for real world applications with a third, **unknown** value to reflect the fact that the value of a variable may not be known. Note that **unknown** can be interpreted as "either **true** or **false**", i.e.

$$\textbf{unknown} = \textbf{true} \lor \textbf{false}$$

The result of any logical operation with any of its operand being **unknown** is most often, but not always **unknown**, i.e. an additional column and row is added to the operation tables with all the values being **unknown** in them.

The following Table 2.3 shows the extended operation table for the logical **or** operation. It is seen from the second row and second column of the table that the logical value **true** in any of the operands will "improve" the uncertainty given by the **unknown** value of the other operand.

*Table 2.3.* Operation table of the extended "or" operation

| $a \vee b$<br>$a \downarrow \quad b \rightarrow$ | false | true | unknown |
|---|---|---|---|
| false | false | true | unknown |
| true | true | true | true |
| unknown | unknown | true | unknown |

## 2.2 SYNTAX AND SEMANTICS OF RULES

A rule is nothing else but a conditional statement, i.e. an

**"if...then..."**

statement. The syntax of a rule consists of the following elements.

1. *Predicates*
   Predicates are elementary logical sentences, their value can be any of the set

   $$\{\textbf{true}, \textbf{false}, \textbf{unknown}\}$$

   They usually contain arithmetic relations ( $=$ , $\neq$ , $\leq$ , $>$ , $<$ ) and they may contain qualitative or symbolic constants (e.g. **low**, **high**, **very small**, **open** etc.).

   Simple examples of predicates from an intelligent control system are:

   $$p_1 = (\kappa = \textbf{on}) \;\; ; \;\; p_2 = (T < 300) \;\; ; \;\; p_3 = (h = \textbf{low})$$

   $$p_4 = (error = \text{"}tank\ overflow\text{"})$$

   where $p_1$, $p_2$ and $p_3$ are arithmetic predicates. The variables in the predicates, $T$ being a temperature, $\kappa$ an on–off switch and $h$ a level, are measured signals, that is, time–varying variables. If, for example, temperature $T$ in a given time instance is equal to 350 $^oK$ then predicate $p_2$ above is **false**.

   It is important to emphasize that *the value of predicates depending on measured signals is time-dependent, that is this value is also a (logical valued) signal in itself.*

2. *Logical expressions*
   A logical expression contains:

   - atomic formulas which can be either predicates or logical variables or logical constants (i.e. **true**, **false** or **unknown**),

- logical operations ($\neg$, $\wedge$, $\vee$, $\rightarrow$)

and obeys the syntax rules of mathematical logics.

3. *Rules*
A rule is in the following syntactical form:

$$\textbf{if} \ \ \textit{condition} \ \ \textbf{then} \ \ \textit{consequence};$$

where *condition* and *consequence* are logical expressions. An equivalent syntactical form of the rule above is in the form of an implication:

$$\textit{condition} \ \ \rightarrow \ \ \textit{consequence}$$

Note that a rule is a logical expression itself.

The *semantics of a rule*, i.e. its meaning when we use it, depends on the goal of the reasoning. Normally, the logical expression *condition* is checked first to see if it is **true**, using the values of the predicates. If this is the case then the rule can be applied or executed (the rule "fires"). When *applying or executing a rule* its *consequence* is *made* **true** by changing the value of the corresponding predicates.

---

**EXAMPLE 2.3** A simple rule set

Consider a simple rule set defined on the following set of predicates:

$$P = \{p_1, p_2, p_3, p_4\} \tag{2.2}$$

$$\textbf{if} \ \ (p_1 \ \textbf{and} \ p_2) \ \ \textbf{then} \ \ p_3; \tag{2.3}$$

$$\textbf{if} \ \ (p_3 \ \textbf{and} \ p_4) \ \ \textbf{then} \ \ p_1; \tag{2.4}$$

The equivalent implication form of the rule set above is

$$(p_1 \wedge p_2) \ \ \rightarrow \ \ p_3;$$

$$(p_3 \wedge p_4) \ \ \rightarrow \ \ p_1;$$

---

## 2.3 DATALOG RULE SETS

There is a simple special case of rule sets called *datalog rule set* which has a nice and transparent structure and advantageous mathematical as well as computational properties [14]. A rule set should possess the following properties to qualify as a datalog rule set.

D1: *There is no function symbol in the arguments of the rules' predicates.*

D2: *There is no negation* $\neg$ *applied to the predicates and the rules are in the following form:*

$$(p_{i_1} \wedge \cdots \wedge p_{i_n}) \quad \rightarrow \quad q_i; \tag{2.5}$$

where $p_{i_1}, \ldots, p_{i_n}$ and $q_i$ are predicates.

D3: *The rules should be "safe rules", that is their value should be evaluated in finite number of steps.* This requirement implies that the range space of any of the variables in the arguments of the rules should be finite.

The rule set of an intelligent control system is almost always in datalog form or if it is not, then can easily be transformed into that form with the following manipulations and considerations.

M1: *Remove function symbols* for requirement D1.
In order to understand why we should avoid rules with function symbols in their predicates' arguments, we recall that most of the special symbols such as *sin* or *exp* are computed by summing the terms in their Taylor series expansion. This may require - at least theoretically - an infinite number of computational steps to be performed to achieve a given precision.

One may introduce new variables which can be pre-computed containing the function symbols present in the argument of a rule's predicate.

M2: *Remove negations and disjunctions ($\neg$ and $\vee$ operations)* for requirement D2.
Disjunctions in the *condition* can be removed by transforming the rule as a logical expression into its implicative normal form. Then in the condition part only conjunctions ($\wedge$ operations) and negations and in the consequence part only disjunctions ($\vee$ operations) and negations remain.

Thereafter we can see that we most often have arithmetic predicates in the rules of an intelligent control system where we can perform the *negation of the arithmetic relation* present in the predicate such as:

$$\neg(a > b) = (a \leq b) \quad , \quad \neg(a = b) = (a \neq b) \quad etc.$$

Thus we can get rid of the negations.

The only property which remains is the existence of a single predicate in each of the rules (2.5). This can be ensured by multiplying the rules with their disjunction in their *consequence* part in the following way:

$$(s_i) : \quad (p_{i_1} \wedge \cdots \wedge p_{i_n}) \quad \rightarrow \quad (q_{i_1} \vee \cdots \vee q_{i_m});$$

*becomes*

$$(s_{i_1}) : \quad (p_{i_1} \wedge \cdots \wedge p_{i_n}) \quad \rightarrow \quad q_{i_1};$$

$$\cdots$$

$$(s_{i_n}) : \quad (p_{i_1} \wedge \cdots \wedge p_{i_n}) \quad \rightarrow \quad q_{i_m};$$

M3: *Consider the finite digit realization of real numbers in computer controlled systems* for requirement D3.

### 2.3.1 THE DEPENDENCE GRAPH OF DATALOG RULE SETS

Datalog rule sets have important properties from the viewpoint of their analysis and execution (reasoning). Their structure can be conveniently described by the so called *dependence graph.*

The dependence graph $D = (V_D, E_D)$ of a datalog rule set is a directed graph which is constructed by the following steps.

1. The vertex set of the graph is the set of the predicates in the rule set, i.e.
$$V_D = P$$
2. Two vertices $p_i$ and $p_j$ are connected by a directed edge $(p_i, p_j) \in E_D$ if there is a rule in the rule set such that $p_i$ is present in the *condition* part and $p_j$ is the *consequence*.
3. We may label the edges $(p_i, p_j)$ by the rule identifier they originate from.

Observe that a rule from the rule set gives rise to as many edges as many predicates are in its *condition* part. All edges originating from the same rule terminate at the same predicate vertex, which is the *consequence* of the rule.

The dependence graph gives information about how the predicate values depend on each other. The following properties of the dependence graph are important from the viewpoint of executing of the rule set, that is, from the viewpoint of reasoning:

- The *set of entrances of the dependence graph*, that is the set of edges with no inward directed edges are the *root predicates* of the set. Their values should be given if we want to compute the value of the other predicates.

- *Directed circles* show that the dependence between the values of the predicates in the circle is not unique: the result of the computation may depend on the computation order.

*If there is no directed circle in the dependence graph of a datalog rule set then we obtain the same reasoning (evaluation) result regardless of the computation order.*

The following example shows a simple dependence graph.

---

**EXAMPLE 2.4** Dependence graph of a simple rule set

Consider a simple rule set defined on the following set of predicates:

$$P = \{p_1, p_2, p_3, p_4\} \tag{2.6}$$

The implication form of the rule set is assumed to be

$$\begin{aligned} (s1) &: \quad (p_1 \wedge p_2) \rightarrow p_3; \\ (s2) &: \quad (p_3 \wedge p_4) \rightarrow p_1; \end{aligned}$$

Note that this is the same rule set as in Example 2.3.

The dependence graph of the rule set is shown in Fig. 2.1. The edges are labeled by the rule identifier they come from. It can be seen that there is a circle joining the vertices $(p_1, p_3)$ on the dependence graph.

---

## 3. OBJECTS

Object-oriented languages, like C++ are quite common in all application areas not only in intelligent software systems [15]. Some of their properties, however, are excellent for knowledge based systems therefore this section contains a brief summary of object-oriented software systems from the viewpoint of intelligent control applications.

The things or items in the focus of our attention are abstract *objects*. Objects can be classified into abstract classes according to the properties

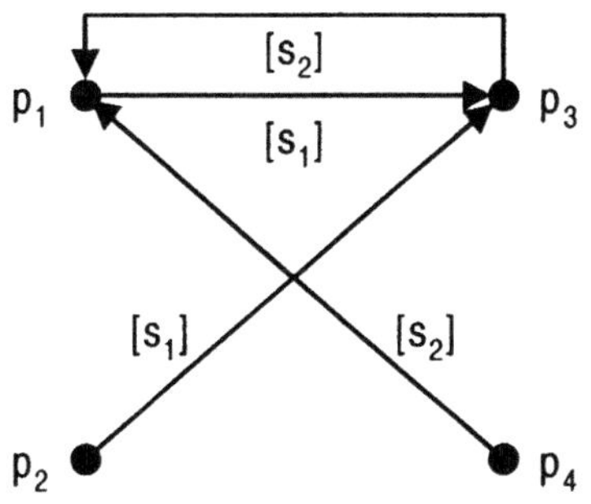


*Figure 2.1.* Dependence graph of a rule set

they have in common. The common properties are *attributes* of the class while the objects as entities of a class may have their own individual properties.

This understanding of a class makes it possible to use a class as a general knowledge element, which has both passive (data-like) and active (procedural) attributes associated with it. This way the description does not only contain the description of the knowledge element itself but also that of its behaviour. Any concrete object then belongs to a class as its entity.

Classes form so called *class hierarchies*, where sub-classes inherit their data and procedural attributes from their parent class or super-class. The class hierarchies are organized in such a way that the parent class of a given class is unique, therefore the hierarchy structure is given by a tree (a graph with no circles).

The descriptions of classes are put into the declaration part of a program.

A simple example shows how the declaration of a simple object may look like.

---

**EXAMPLE 2.5** A simple class declaration

Let us consider a simple tube equipped with a valve to open or close the flow going through the tube. Measurement devices for measuring the key thermodynamical properties of the flow, that is, the temperature ($T$) and the flowrate $v$ are also assumed to be present.

The following declaration frame indicates how these knowledge elements and some of their behaviour can be represented as attributes and procedures of a "tube" object.

```
{ class head  }  class       tube
{ attributes  }  val:        valve;
                 T,v:        measurement-device;
{ procedure   }  procedure   open-valve   (error-code);
                 . . .       { statements to open }
                 end         {open-valve}
{ class body  }  . . .       { statements to initialize }
                 end;        { tube }
```

Observe, that the equipments belonging to the tube are described in the form of attributes, and these are objects of different types: "val" being a valve, and "T" and "v" measurement-devices.

There is only one procedure defined for opening the valve (the own valve of the tube!) named "open-valve".

---

The main properties of object-oriented tools explain their widespread use in knowledge based systems.

1. *Instances can be created from a class* by suitable parametrization. The instances become individual objects of their own.

   In the simple example above we can create two different instances of the equipped tube described by "class tube" if we write the following in the executable part of our code:

   ```
   tube-one:= new tube;
   tube-two:= new tube;
   ```

2. *Objects are encapsulated*, which means that they have their "private life", their properties can only be changed by calling their procedures. Thus one can reach the attributes of an object only via its own procedures.

   If we take the simple example of the tube above (Example 2.5) again, then we can open the valve attached to the second tube if we write:

   ```
   tube-two.open(err-code-2);
   ```

Then this valve will be open, but the valve attached to "tube-one" remains in its previous state.

3. *The properties of a parent class are inherited to its sub-classes.* A parent class is in a so called **is_a** relation with its sub-class. (See later in section 5. of this Chapter on semantic nets.) Class hierarchies can also be constructed.

The following simple example shows a possible class hierarchy for the coffee machine, which is described in Appendix B.

---

**EXAMPLE 2.6** A simple class hierarchy

Consider again the tube example described above in Example 2.5, but now with different tubes. Assume, we have a basic tube type with only one valve attached to it, and an "advanced" tube type, where measurement devices are also present. In order to be able to describe instances of both types sharing common attributes and behaviour, we construct the following class hierarchy in the declaration part of our program.

```
{parent class   }  class tube
{p-attributes   }  val:          valve;
{p-procedure    }  procedure     open-valve  (error-code);
                   ...           {statements to open}
                   end           {open-valve}
{p-class body   }  ...           {statements to initialize}
                   end;          {tube}

{sub-class     }  tube class meas-tube
{s-attributes  }  T,v:                    measurement-device;
{s-procedure   }  procedure               measure   (value);
                  ...                     {statements to get the value}
                  end                     {measure}
{s-class body  }  ...                     {statements to initialize}
                  end;                    {meas-tube}
```

---

# 4. FRAMES

Frames [16] are knowledge structures with special pre-defined knowledge elements connected by semantic relationships.

Frames can be seen as extensions of records with standard active elements [17]. On the other hand, frames are similar to objects in the sense that instances can be generated from them and they can also form frame hierarchies with inheritance. The properties of frames above explain why they are convenient for knowledge representation.

Frames as elementary knowledge structures have the following standard parts.

- *Slots*

  Slots play the same role in a frame as fields in a record. The attributes of a slot are its identifier (or name), type and value. In order to make knowledge representation easier, the type declaration for slots is more flexible and can be changed during run-time.

  The following simple example, a part of a declaration in a frame-based environment, illustrates the flexibility of the type declaration.

```
measured-data frame;
              value:    real or byte;
              status:   byte;
end           {measured-data};
```

- *Daemons*

  Daemons are standard built-in procedures provided for each slot. They are automatically invoked when a predefined change in the value of the slot is taking place.

  The usual daemons are as follows.

  - `if_added` contains the actions to be performed when the slot gets its first non-`nil` value;
  - `if_removed` is the procedure to be executed when the value of the slot is deleted (becomes `nil`);
  - `if_needed` describes the steps to be performed when the value of the slot is read (retrieved);
  - `if_changed` is the daemon which is invoked when the value of the slot is changed.

The *use of frames* resembles the use of objects. The main difference is that the number and role of the procedures, defined for a frame are fixed and built-in by the frame environment. Of course, the user determines the executable part of the daemons and it may even be empty.

It is important to note that one can change the value of slot in any frame instance. This way daemons can invoke (or call) each other via changing slot values in their procedure bodies.

Similarly to an object-oriented environment, frames define types of knowledge elements the same way as classes do. Their definition is in the declaration part of the program. Frame hierarchies connected by inheritance can also be formed. Any number of instances can be created from any frame in the executable part of the program.

The properties of a frame environment can be summarized as follows.

1. A frame system contains both passive ingredients in the slot values and active elements in the executable parts of the daemons.

2. The operation of a frame system is described in an indirect way. It is embedded in the daemons of the frame instances in the frame system.

In conclusion: *frame-based knowledge representation is flexible but it is difficult to see through, verify and validate.*

## 5. SEMANTIC NETS

Semantic nets are graphic tools for describing semantic relationships between knowledge items in a knowledge base. The properties and relationships of the knowledge objects and classes are described by a *directed graph.* The vertices of the graph correspond to the objects and their attributes or properties: the labelled edges depict the relationships between the vertices.

Most of the relationships in a semantic net fall into pre-defined categories. The most common relationships are as follows.

- `is_a`
  which means that *objectA* is an instance of *objectB* if the relationship

  *objectA* `is_a` *objectB*

  holds.

- `part_of`
  meaning that *objectA* is a part of or an attribute of *objectB* when

  *objectA* `part_of` *objectB*

  holds.

Observe, that the relationships above are necessary and sufficient to describe the relationships in an object-oriented knowledge base. Other knowledge representation methods, such as frames, may call for other pre-defined relationship categories.

The real semantic relationships are strongly problem or knowledge base dependent, therefore cannot be given in advance.

It is important to note that semantic relationships can also be described by binary relations. Thus the following expressions are equivalent but they are in different syntactical forms:

$$\text{"}objectA\ \texttt{part\_of}\ objectB\text{"} \equiv \text{"}\texttt{part\_of}(objectA, objectB)\text{"}$$

$$\text{"}objectA\ \texttt{is\_a}\ objectB\text{"} \equiv \text{"}\texttt{is\_a}(objectA, objectB)\text{"}$$

Fig. 2.2 shows how different relationships are depicted in a semantic net. The following semantic relationships are depicted:

*Mike* is_a *teacher*

*table* part_of *room*

*flower* colour *blue*

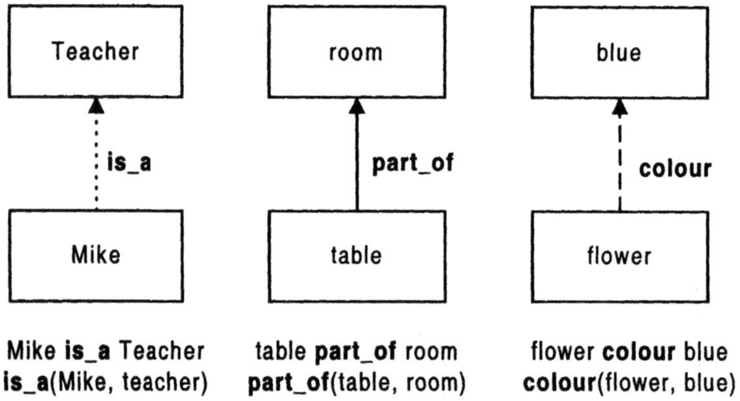


*Figure 2.2.* Simple semantic relations

Semantic nets are *meta-knowledge structures* because they describe knowledge about knowledge items in a knowledge base. They can be used together with any type of knowledge representation method. They show the structure of a knowledge base.

In summary: *semantic nets are mainly used for knowledge base verification, validation and diagnostic purposes.*

---

**EXAMPLE 2.7** A simple semantic net

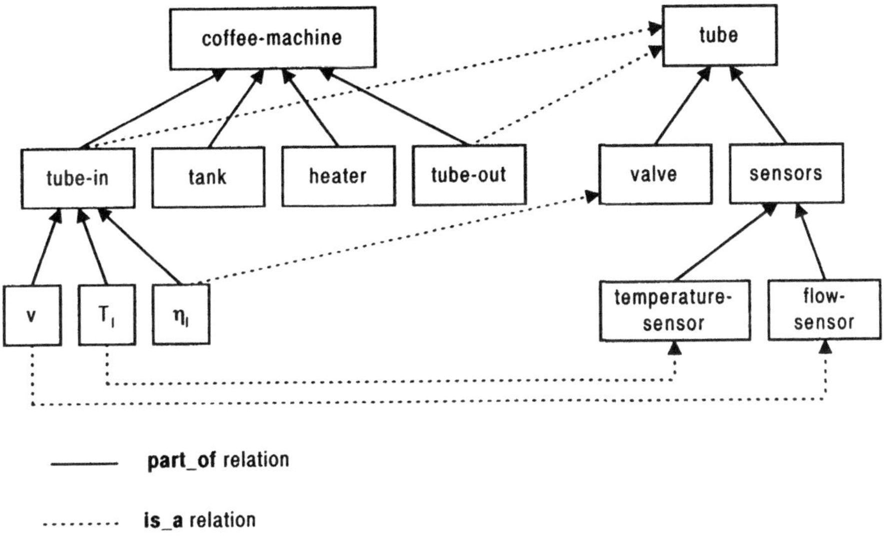


*Figure 2.3.* A simple semantic net

Fig. 2.3 shows part of the semantic net that describes the objects and their connections in a model of the coffee machine shown in Fig. B.1 in Appendix B.

---

Chapter 3

# REASONING AND SEARCH IN RULE-BASED EXPERT SYSTEMS

The basic methods of reasoning are described and the close connection between reasoning and search is explained in the following sections of this chapter:

- Solving problems by reasoning [18] - [21]
- Forward chaining [20], [21], [4]-[8]
- Backward chaining [20], [21], [4]-[8]
- Search methods and heuristics [4]-[8]

## 1. SOLVING PROBLEMS BY REASONING

The fundamental architecture of an expert system has already been discussed in section 2. of Chapter 1. The main components and their connections have also been depicted there in Fig. 1.1. An expert system consists of the following components:

- a *knowledge base* that contains expert knowledge in some specific domain
- an *inference engine* that manipulates the knowledge base to find answers for given problems
- a *user interface* that helps the system to communicate with the user
- a *knowledge base maintenance system* that fills, modifies and analyzes the knowledge base
- a *developers' interface* that helps the system to communicate with the knowledge engineer

## 1.1 THE STRUCTURE OF THE KNOWLEDGE BASE

The knowledge base of a rule-based expert system consists of two parts:

- The *facts* or *predicates* represent declarative knowledge about the units or sets of the given problem. They are statements with either true or false values, in extended cases they may take other discrete values such as unknown. The value of a predicate can change in time and also during reasoning.

- *Connections* or *rules* are used to represent heuristics or "rules of thumb", which typically specify actions that may be taken in a given situation. They are operated by the inference engine to modify the facts. These rules can only be changed by the knowledge engineer during knowledge base maintenance.

The syntax and semantics of rules have already been discussed in section 2.2 in Chapter 2.

At any given time the state of the knowledge base is the value of all the predicates, which can be represented by a state vector.

$$\underline{a} = \begin{bmatrix} p_1 \\ . \\ . \\ . \\ p_{n_P} \end{bmatrix}$$

where
$p_i = \{t(true),\ f(false),\ u(unknown)\}$ and
$n_P$ is the number of the predicates.

The set of all states of the knowledge base that can be reached from the initial state (or from a set of possible initial states) by any sequence of actions, including the initial and terminal states are contained in the *state-space*.

The rules consist of a *condition* or *premise*, which tests the logical value of a set of facts at every stage of the reasoning process followed by an *action* or *consequence* describing what to do when the rule fires.

**if** *condition* **then** *action*

Both the condition and the consequence part of a rule represent statements which consist of disjunctions or conjunctions of facts. For the sake of simplicity *datalog rules* are used in this chapter where the condition part contains a conjunction of predicates and there's only one predicate in the action part.

$$(r_i) : (p_{i_1} \wedge \cdots \wedge p_{in_i}) \rightarrow q_i$$

For more about datalog rules, see section 2.3 in Chapter 2.

For the purpose of analysis, a special data structure is constructed to describe such a rule-base:

$$\begin{aligned} rule\_base &= \{ \ (n_P \ , \ n_R) \\ (r_1) &: \ (p_{1_1} \wedge \cdots \wedge p_{1_{n_1}}) \ \rightarrow \ q_1; \\ \ldots & \quad \ldots\ldots\ldots\ldots\ldots \\ (r_{n_R}) &: \ (p_{n_{R1}} \wedge \cdots \wedge p_{n_{Rn_R}}) \ \rightarrow \ q_{n_R}; \\ & \quad \} \end{aligned} \tag{3.1}$$

where $n_P$ is the number of predicates and $n_R$ is the number of rules.

## 1.2 THE REASONING ALGORITHM

Rules are used by the inference engine in order to derive new knowledge or information. An elementary reasoning step applies a single rule and consists of the following sub-steps:

- selecting one of the applicable rules (a rule is applicable when the predicates in its condition part are true)

  The inference engine *matches* facts with the condition of the rules to determine which rules should be applied and *selects* the most appropriate rule.

- modifying the facts by the selected rule (the logical value of the predicates in the action (conclusion) part of the rule is set to true)

  The selected rule is *fired* by the inference engine and the action associated with it is *executed.*

The inference engine repeats this elementary reasoning step in a loop through all the rules and facts until no more conclusion can be reached or the termination conditions are satisfied. (see in Fig. 3.1)

It is important to note that new facts can be deduced during *reasoning* from the existing facts. The reasoning tool is the application of rules or in other words the matching of rules. The aim of reasoning is to reach (construct) a goal state or prove a goal statement. The basic

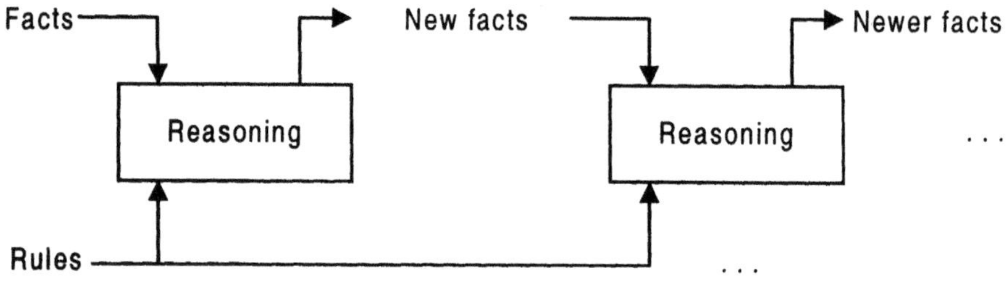


*Figure 3.1.* The steps of reasoning

mathematical formula used in the reasoning is the famous *modus ponens* in the following form:

$$\frac{\begin{array}{c} A \\ A \rightarrow B \end{array}}{B} \quad or \quad A,\ A \rightarrow B \Rightarrow B$$

or

If A is true and B follows from A, then B is true.

Modus ponens can be used in two ways. Reasoning can be started with the facts in the knowledge base, in which case modus ponens generates new conclusions that in the next turn allow more inferences to be made. This is called *forward reasoning*. Alternatively, reasoning can be started with something to be proved. In this case we look for an implication with its consequence part containing the predicate to be proved. Thereafter we prove the predicates in the condition part of this implication. This is called *backward reasoning*, because it uses modus ponens backward. In case of both directions a reasoning path, that is a chain of rules can be constructed between the facts and the goal state.

This reasoning chain can be seen as a path in the state-space, a sequence of rules leading from one state to another. Problem solving (reaching any goal state from the initial state) is performed by applying the rules one after the other expounded on the state-space. This view on reasoning can be illustrated on the state-space of the knowledge base where the actual state is moved by the rules during reasoning. These movements are performed by only one co-ordinate direction at a time instance in case of datalog rules.

The sequences of reasoning steps correspond to a *graph traversal* from an initial state to one or more possible, acceptable or optimal goal states.

This way a reasoning problem can be formulated as a *searching problem* in the state-space where rules are assigned to the possible actions. In this context, search is a general purpose method to solve problems where the initial state, the actions and a goal state or goal test are given. The aim is to get to a goal state from the start state via a series of successor states. The solution path from the initial state to a state satisfying the goal test consists of transitions from a state to another state executed one after another.

---

**EXAMPLE 3.1** Reasoning in the state-space

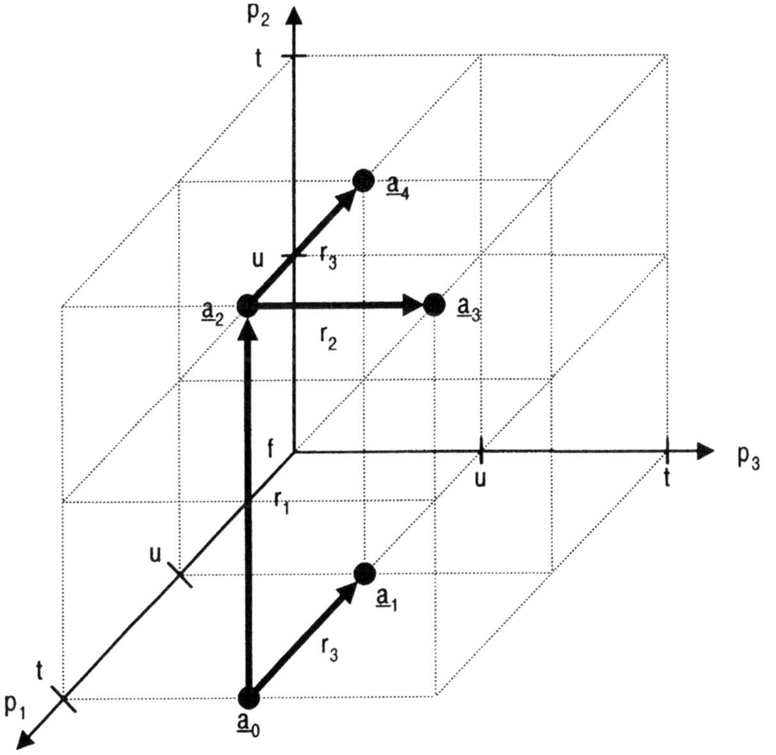


*Figure 3.2.* Reasoning in the state-space

Let us define the initial state and the rules as follows:

$$\underline{a}_0 = \begin{bmatrix} t \\ f \\ u \end{bmatrix}$$

where $t$ denotes **true**, $f$ is **false** and $u$ is **unknown**.

$$(r_1) \quad : \quad \textbf{if } p_1 = t \textbf{ then } p_2 = t$$
$$(r_2) \quad : \quad \textbf{if } p_2 = t \textbf{ then } p_3 = t$$
$$(r_3) \quad : \quad \textbf{if } p_3 = u \textbf{ then } p_1 = u$$

Reasoning in the state-space is illustrated in Fig. 3.2.

There are two applicable rules, namely $r_1$ and $r_3$ in the initial state. State $\underline{a}_1$ (which is a terminal state, that is no rule can be applied) is reached by rule $r_3$. In state $\underline{a}_2$ there are again two applicable rules $r_2$ and $r_3$, this state is reached by rule $r_1$.

---

At any given time there can be many applicable rules matching the facts and the result of reasoning could depend on the order of their application. This situation is called a *conflict* and it is represented by a *branch* of the search tree in the state-space. The number of branches is equal to the number of applicable rules in a state. Choosing which rule to apply next is called a *conflict resolution*.

A directed search graph in the state-space is defined by the rule set and the initial state. In this graph each node represents a state of the state-space and each arc represents an action changing the state to another. This search graph in the state-space is not given explicitly in the beginning of the reasoning process, but is exhibited gradually as the rules take a node in the state-space as input and produce its successors. So the graph is given in an implicit way, and it is generated during reasoning (it is generated on the fly).
Fig. 3.3 shows that the search graph in the state-space can be transformed into a two-dimensional graph preserving the adjacency relations.

It is emphasized again that only a local part of the graph can be seen at a given state, namely the nodes which have been traversed earlier and the branches of the node. With this local information we need to decide where the goal node may be, which way we prefer to reach it and how to traverse the graph.

## 1.3 CONFLICT RESOLUTION

For the majority of problems, there is no exact solution strategy optimal to every possible reasoning task. Moreover, it is not an excellent

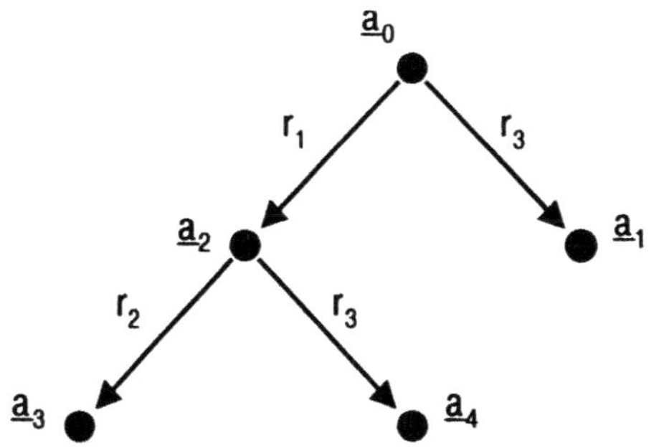


*Figure 3.3.* The reasoning graph

idea to solve the problems by testing every possible way of solution because of the combinatorical explosion. Even for most of the real practical problems, there is no need to produce all possible solutions, the aim is to obtain a "good enough" solution in a "short enough" time.

*Conflict resolution* aims at choosing which rule to apply next from the applicable ones. It is the most important algorithm of the inference engine. It almost always contains *heuristic knowledge*, that is extra knowledge beyond the state-space, which can be regarded as metaknowledge about the structure of the rule-base.

The notion of heuristics has no exact definition, but all heuristic procedures exhibit two significant properties:

- A "good enough" solution is found in most cases, but the optimal solution or any solution is not guaranteed.
- Heuristic procedures considerably improve the efficiency of problem solving by reducing the number of attempts to reach the solution.

The function of heuristics is to determine the order in which to apply rules during reasoning. Heuristics may be very simple or quite complex. A good heuristics can be characterized by the following properties:

- It is used and computed efficiently.
- It is a good estimate, but it does not overestimate the effective costs.

The most widely used methods of conflict resolution are as follows:

- using the first applicable rule (when the rules are placed in order of importance),
- assigning priority to rules,
- using other heuristic methods.

## 1.4 EXPLANATION OF THE REASONING

The ability of an expert system to explain its reasoning is one of the most powerful attributes. Since the system remembers its logical chain of reasoning, it is able to explain how it arrived at a conclusion whenever the user asks for an explanation.

The explanation can give information about "How?" and "Why?" by tracing the reasoning process. Hypothetical reasoning can also be applied with tracing to answer "What if?" type questions. For more about the explanation facilities provided by an expert system shell see in section 3. in Chapter 5.

# 2. FORWARD REASONING

The simplest reasoning method is *forward reasoning, forward chaining* or *data-driven chaining*. It is used to infer solutions from knowledge that exists in the knowledge base.

## 2.1 THE METHOD OF FORWARD REASONING

Forward reasoning begins with a set of known facts, derives new facts using rules whose conditions or premises match the known facts and continues this process until a goal state is reached or until no further rules have conditions that match the known or derived facts. (see in Fig. 3.4)

The problem of forward reasoning is defined as a standard algorithmic problem as follows.

FORWARD REASONING WITH DEFINED GOAL

*Given:*

- the initial state of fact-base ($\underline{a}_0$)
- the rule-base
- a goal state or goal states of fact-base ($\underline{a}_g$)

*Question:*
Is $\underline{a}_g$ a consequence of $\underline{a}_0$?
(Can $\underline{a}_g$ be derived from $\underline{a}_0$ by the rules?)

The above problem is a decision problem where the whole search tree must be traversed in the worst case to get an answer to the question. As the size of the tree (the number of nodes) increases the number of computational steps exponentially, the problem is NP-complete.

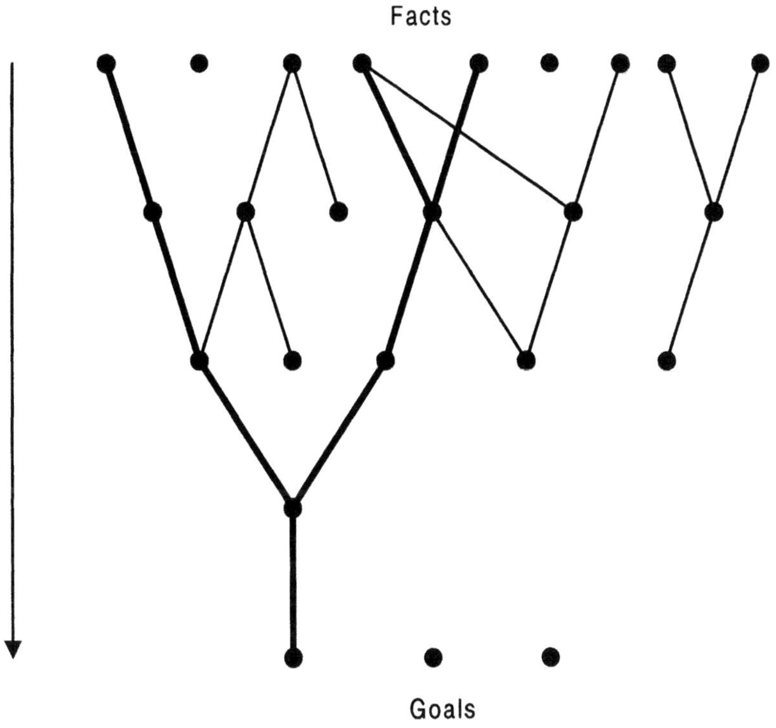


*Figure 3.4.* The forward reasoning

A search variant of the problem above is obtained if we do not specify the goal state.

FORWARD REASONING

*Given:*

- the initial state of fact-base ($\underline{a}_0$)
- the rule-base

*Compute:*
all the possible consequences of the initial state(s).

This is a search problem, where again, the NP-completeness follows from the problem specification.

In forward chaining the search graph in the state-space is built from the initial state $\underline{a}_0$. During the traversal of the graph the condition parts of rules are matched to the fact-base and one of the applicable rules is executed, that is, the facts in the consequence part of the selected rule are added to or some facts are deleted from the fact-base. With the application of the rule we can get to the next state. If this state is one of the goal states of the FORWARD REASONING WITH DEFINED GOAL problem, then the algorithm terminates.

If there is no more applicable rule and the terminal state is not in the goal state set $\underline{a}_g$, then the algorithm must go back to a state with more applicable rules and should use the next one. The terminal state is observed before stepping back in the case of the FORWARD REASONING WITH DEFINED GOAL problem, where there is no goal state specified. This "going back" described above is called *backtrack* . The backtrack mechanism will try all of the possible rules selecting the first alternative at each state and backtracking to the next alternative when it has pursued all of the paths from the first choice.

The backtrack mechanism that can be applied to the reasoning graph in Fig. 3.2 is illustrated in Fig. 3.5

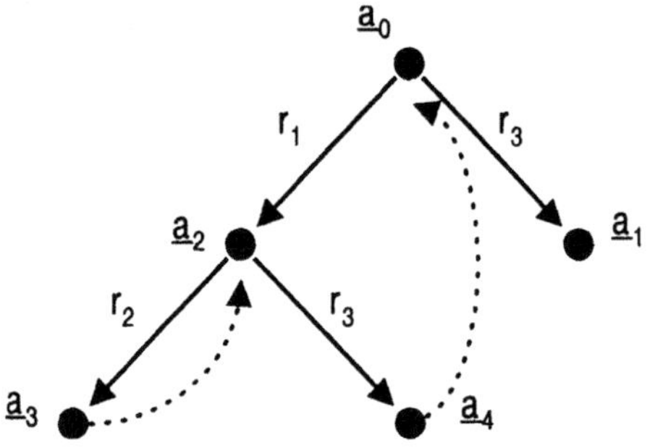


*Figure 3.5.* **The backtrack mechanism on the reasoning graph in Fig. 3.2**

It is important to note that the possible branching alternatives, that is the rules not being examined must be stored in the backtrack mechanism. Therefore, *the whole knowledge base must be locked during reasoning in order to ensure its consistency for the ongoing reasoning process.*

Forward reasoning is recommended for the solution of the following types of problems:

- when all or most of the data are given in the specification of the initial state

  *For example*: the possible minerals of a given region are deduced from geological tests.

- there are several possible goal states, but the information is only used by some resolution paths

*For example:*

- the composition of organic compounds is determined using knowledge gained from different measurements.
- predictions are computed from measured data in a real-time expert system

## 2.2 A SIMPLE CASE STUDY OF FORWARD REASONING

Let us define the initial state of the fact-base as follows:

$$\underline{a}_0 = \begin{bmatrix} A = t \\ B = t \\ C = t \\ D = f \\ E = t \\ F = f \\ G = t \\ H = t \\ Z = f \end{bmatrix}$$

Consider a simple rule set arranged in the order of the priority of rules in order to apply this heuristic for conflict resolution.

$(r_1) : F \wedge B \rightarrow Z$
$(r_2) : G \wedge H \rightarrow \neg C$
$(r_3) : C \wedge D \rightarrow F$
$(r_4) : A \rightarrow D$

Assume that predicate $Z$ is true in goal state $\underline{a}_g$ of the fact-base and the value of the other predicates is indifferent with respect to the goal.

*Question:*
Can goal state $\underline{a}_g$, when $Z$ is true, be derived from $\underline{a}_0$ by the rules?

We will assume that each time the set of rules is tested against the fact-base, only the rules producing a new state of the fact-base are executed.

*Solution:*
Given the above facts and rules, the steps of forward reasoning are as follows (Fig. 3.6):

1. The rules that can fire in the initial state are $G \wedge H \rightarrow \neg C$, and $A \rightarrow D$, because their condition parts are **true** ($G$, $H$ and $A$ in the fact-base). Actually, the first rule fires because it has higher priority. As a consequence, $C$ is removed from the fact-base, that is $C$ is set to **false**.

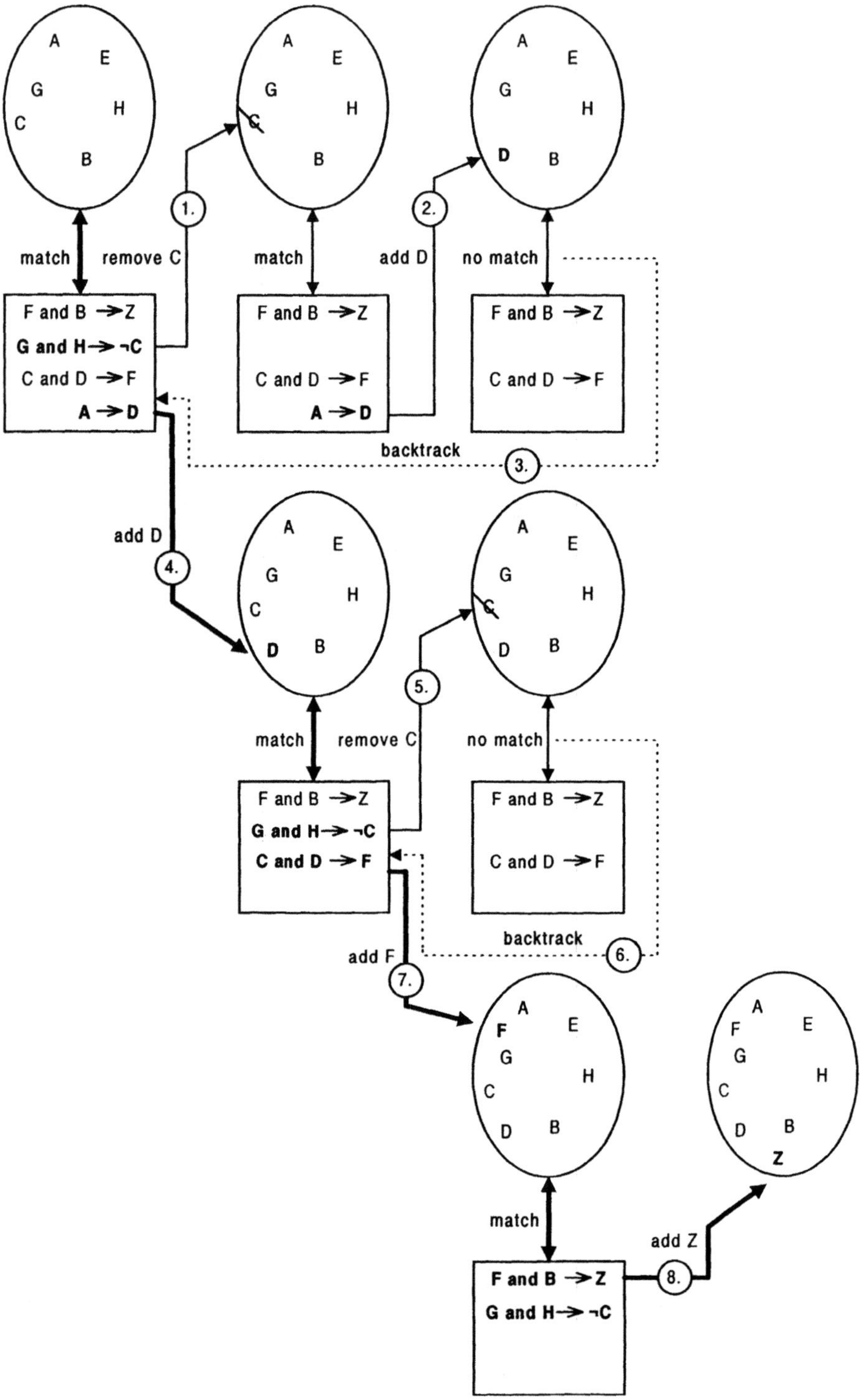


*Figure 3.6.* An example of forward reasoning

2. Then only rule $A \to D$ matches the fact-base in the second step of reasoning. As a result of executing the rule, the existence of $D$ is inferred and $D$ is placed in the fact-base by setting its value to **true**.

3. No rule matching the predicates exists in the resulting state of the fact-base, so we must go back to a preceding state to find more applicable rules.

4. We are again in the initial state and use rule $A \to D$, set the value of $D$ to **true**, that is we add $D$ to the fact-base.

5. The executable rules are $G \wedge H \to \neg C$, and $C \wedge D \to F$. Because of the higher priority the first rule fires, removing $C$ from the fact-base.

6. We need to backtrack again because the rules don't match the predicates of the fact-base.

7. Fact $F$ is inferred and placed in the fact-base as a consequence of rule $C \wedge D \to F$.

8. This in turn causes the first rule $F \wedge B \to Z$ to fire, placing $Z$ in the fact-base. Forward reasoning has succeeded, the goal state is reached, $Z$ is inferred from the initial state.

The inference chain produced by the example in Fig. 3.7 is illustrated in Fig. 3.7.

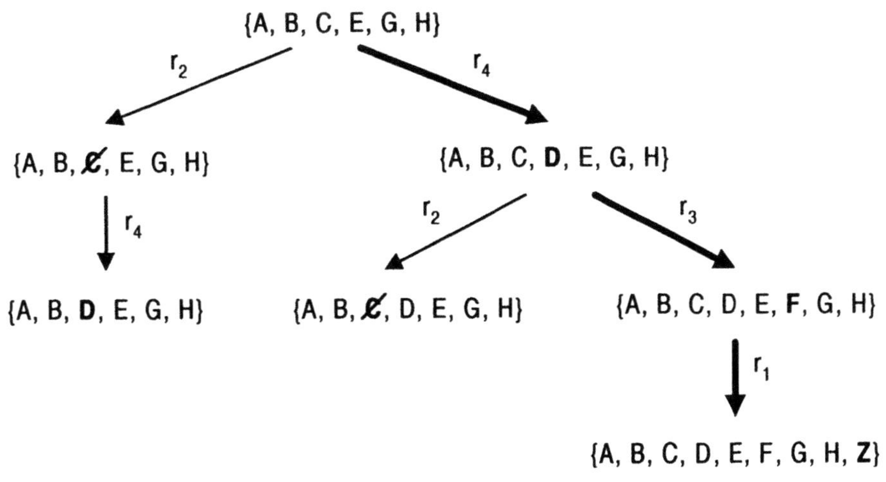


*Figure 3.7.* The inference chain produced by the example in Fig. 3.6

# 3. BACKWARD REASONING

Backward reasoning is applied to infer the causes of a situation, that is the possible facts which lead to a goal state driven by the rules.

Before explaining the backward reasoning technique in detail, a new problem solving method is discussed in this chapter in order to make it easier to understand the method of backward reasoning.

## 3.1 SOLVING PROBLEMS BY REDUCTION

The approach whereby one divides a problem into subproblems and then divides these into further subproblems until there are subproblems that can directly be solved is frequently used in human thinking. The solution of the original problem is traced back to the solution of simple subproblems. This method is called *problem reduction.*

The algorithmic steps of problem reduction are represented by a graph, where the nodes of the graph correspond to the state of problems and the directed edges (or arcs) correspond to the reduction operators splitting the problems into subproblems. The application of a reduction operator could result in more coherent edges from a node. These arcs are called *hyperarcs* and they are connected with circled lines in the figures. The graph containing hyperarcs is called *hypergraph* or *AND-OR graph.*

---

**EXAMPLE 3.2** A simple AND-OR graph

Consider a simple AND-OR graph in Fig. 3.8.

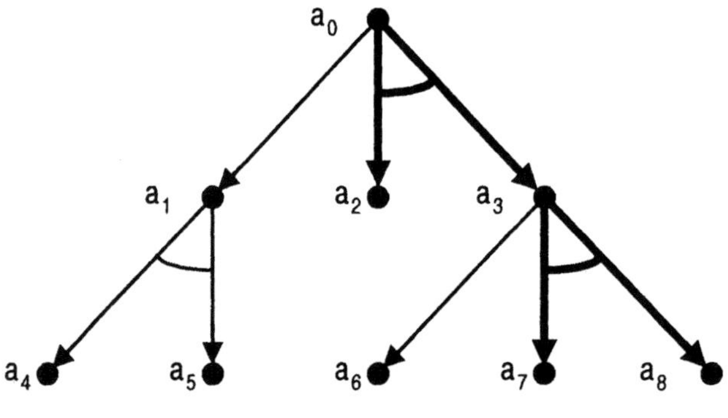


*Figure 3.8.* A simple AND-OR graph

There are two hyperarcs from node $a_0$, one from $a_1$ and two from $a_3$. The hyperarcs from $a_0$ to $a_1$ and from $a_3$ to $a_6$ only contain one common directed arc, but the hyperarcs from $a_0$ to $a_2$ and $a_3$, from $a_1$ to $a_4$ and $a_5$, and from $a_3$ to $a_7$ and $a_8$ consist of two common arcs.

Node $a_0$ has three children nodes ($a_1$, $a_2$ and $a_3$) and there is a narrower, so called *AND* connection between $a_2$ and $a_3$ because they belong to the same hyperarc. Node $a_1$ is connected to them with an *OR* connection.

---

The nodes of the AND-OR graph connected to each other with AND connections represent subproblems of which *all* should be solved. But in case of an OR connection it is enough to solve *one* subproblem.

A *solution* in an AND-OR graph is called *hyperpath*, which is a *subgraph* from the initial node to the set of goal nodes. A possible solution graph is shown in bold in Fig. 3.8.

## 3.2 THE METHOD OF BACKWARD REASONING

The second basic rule-based reasoning strategy is *backward reasoning, backward chaining* or *goal-driven chaining.* In this reasoning strategy we first set the goal as a hypothesis and then we attempt to prove it. (see Fig. 3.9) If it cannot be proved directly from the initial state of the facts, then the goal is broken down into subgoals in each phase of the reasoning process until the conclusion is proved or disproved. The solution of a backward reasoning problem can be conveniently described using an AND-OR graph.

In the backward reasoning strategy, rules are used in a reverse direction, from their action part to the condition part. A rule is able to fire when its action part contains the current subgoal needed to prove. Similarly to forward reasoning problems, backward reasoning problems are defined as follows.

BACKWARD REASONING WITH DEFINED FACTS

*Given:*

- a goal state of the fact-base ($\underline{a}_g$)
- the rule-base
- one or more given states of the fact-base ($\underline{a}_0$)

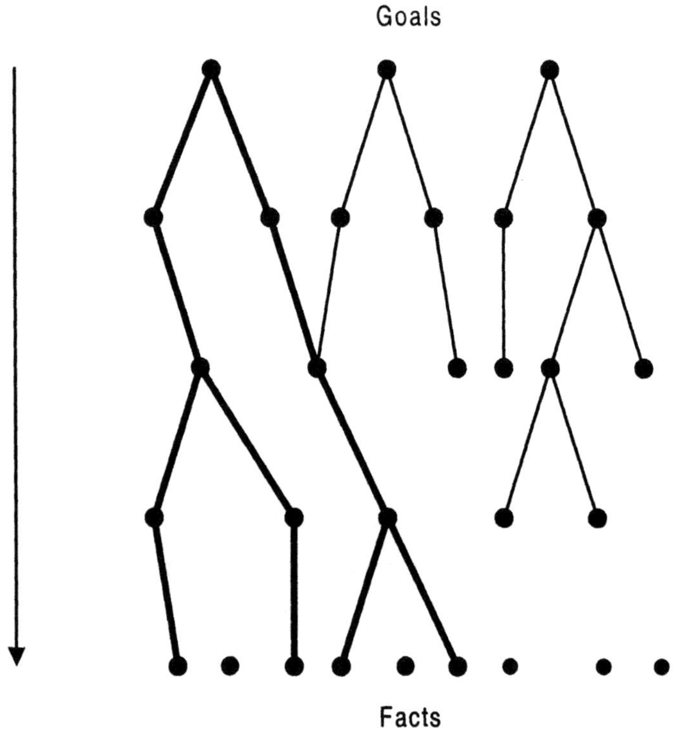


*Figure 3.9.* Backward reasoning

*Question:*
Can $\underline{a}_0$ be a reason of $\underline{a}_g$?
(Can $\underline{a}_g$ be derived from $\underline{a}_0$ by the rules?)

This is a decision task where in the worst case, the whole search tree must be traversed. As the size of the tree (the number of nodes) increases, the number of necessary computation steps increases exponentially, thus the problem is NP-complete.

The search variant of the problem is obtained when no other state is given.

BACKWARD REASONING

*Given:*

- a goal state of the fact-base ($\underline{a}_g$)

- the rule-base

*Compute:*
all of the possible reasons of $\underline{a}_g$

This is a search problem, which is again NP-complete.

In backward reasoning, we start with the goal state (to be proved) of the fact-base ($\underline{a}_g$) and find a rule containing some predicates from $\underline{a}_g$ in its consequence part. The reason of $\underline{a}_g$ may be the facts in the condition part of this rule in the case of a BACKWARD REASONING problem. Otherwise, to find all of the possible reason backward reasoning is accomplished with the predicates in the condition part, which are treated as new subgoals. Besides them the procedure backtrack to the states that have more applicable rules.

In the case of BACKWARD REASONING WITH DEFINED FACTS the algorithm terminates if state $\underline{a}_0$ is reached and all of the subgoals are matched to the fact-base. The procedure backtracks if the proof of any of the subgoals is not succeeded, that is there is no fact or rule matching. In case of backtracking, the test of the subgoal is discarded and a new subgoal used to match, and if there is no matching rule then the procedure backtracks to the previous level, and so on.
It is suggested to use backward reasoning for the solution of problems with the following characteristics :

- The goal is given in the specification of the problem.

  *Example*:

  – proving a theorem in mathematics
  – diagnosis in diagnostic systems

- There are a lot of rules in the knowledge base.

  *Example*: proving a theorem in mathematics

- Problem data are not given but must be generated, retrieved or found during problem solving.

  *Example*:

  – diagnosis in medical diagnostic systems
  – diagnostics and identification in real-time expert systems for control

## 3.3 A SIMPLE CASE STUDY OF BACKWARD REASONING

Let us define the initial state of the fact-base as follows:

$$\underline{a}_0 = \begin{bmatrix} A = t \\ B = t \\ C = t \\ D = f \\ E = f \\ F = f \\ G = t \\ H = t \\ Z = f \end{bmatrix}$$

Consider a simple rule set arranged in order of priority as follows:

$(r_1) : H \wedge E \rightarrow F$
$(r_2) : F \wedge B \rightarrow Z$
$(r_3) : C \wedge D \rightarrow F$
$(r_4) : A \rightarrow D$

Also, let $Z =$ **true** in the goal state.
*Question:*
Can the goal state with $Z =$ **true** be derived from the initial state $\underline{a}_0$ by the rules?
In other words, the aim is to prove the existence of $Z$.

The steps of backward reasoning are illustrated in Fig. 3.10 and are as follows:

1. First of all, the inference engine checks the fact-base for $Z$ and since it fails, it searches for rules that conclude $Z$. The first rule which can fire is $F \wedge B \rightarrow Z$ because $Z$ is in its consequence part. Two subgoals - $F$ and $B$ - must then be established in order to conclude $Z$.

2. $F$ is not in the fact-base but the rules $H \wedge E \rightarrow F$, and $C \wedge D \rightarrow F$ conclude $F$.

3. From the higher priority of the first rule, the system decides that $H$ and $E$ must be established to conclude $F$.

4. $H$ is in the fact-base, so the first subgoal of the rule $H \wedge E \rightarrow F$ is satisfied.

5. The second subgoal is not succeeded, because predicate $E$ is neither in the fact-base, nor in the consequent part of any of the rules.

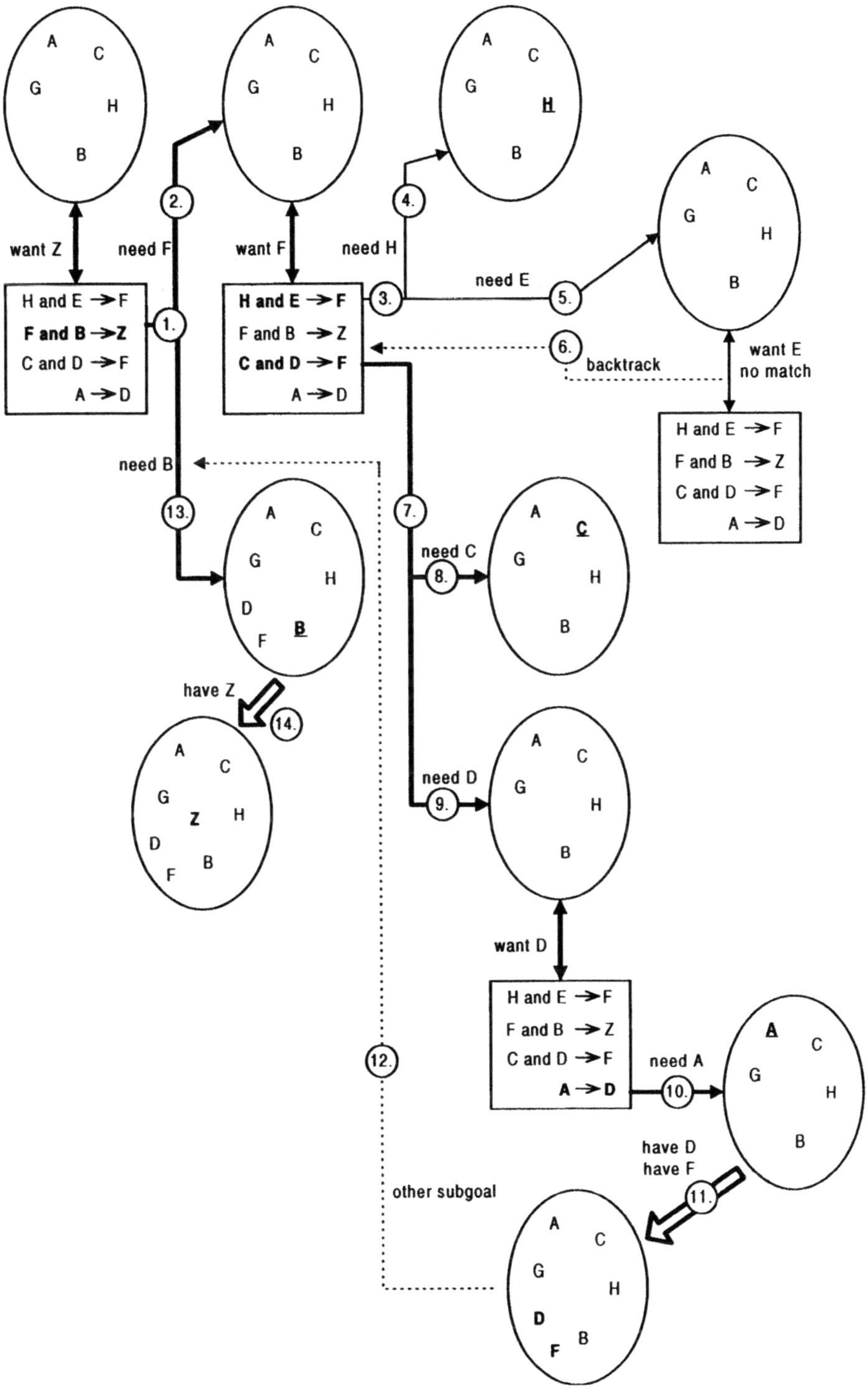


*Figure 3.10.* An example of backward reasoning

6. We need to backtrack to the state mentioned in 2. and use rule $C \wedge D \rightarrow F$.
7. Now we have to establish $C$ and $D$ to conclude $F$.
8. The first subgoal of the rule $C \wedge D \rightarrow F$ is to prove $C$. As $C$ is in the fact-base, it is succeeded.
9. The second subgoal is the verification of $D$. As $D$ is not in the fact-base, we need to find a rule containing predicate $D$ in its consequence part. Rule $A \rightarrow D$ is applicable and the subgoal is to prove $A$.
10. As predicate $A$ is in the fact-base, rule $A \rightarrow D$ is satisfied.
11. Predicate $D$ is established according to rule $A \rightarrow D$ and predicate $F$ is established according to rule $C \wedge D \rightarrow F$ and they are placed in the fact-base.
12. There is still one subgoal unsatisfied: we must prove the existence or the deducibility of predicate $B$ in order to prove $Z$ in rule $F \wedge B \rightarrow Z$.
13. $B$ is in the fact-base, so rule $F \wedge B \rightarrow Z$ is satisfied and $Z$ is put into the fact-base.
14. As $Z$ is in the fact-base and there are no more subgoals, the original goal is established and $Z$ is proved.

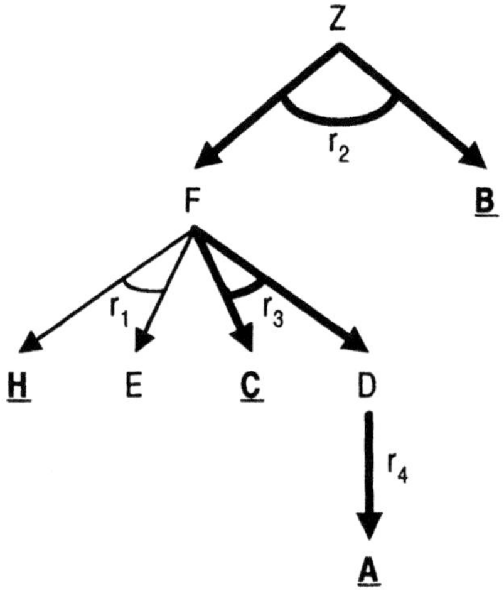


*Figure 3.11.* The AND-OR graph produced by the example in Fig. 3.10

The inference chain produced by the example in Fig. 3.10 is shown in Fig. 3.11.

# 4. BIDIRECTIONAL REASONING

In every special case, the nature of the actual problem determines which reasoning technique is to be applied. However, there may be problems where neither forward chaining nor backward chaining is efficient. If we assume, however, that they operate efficiently at an early stage, it's a good idea to use *bidirectional reasoning* - a combination of backward and forward reasoning. In this reasoning method, the path of rules leading from the start to the goal state are searched from two directions, from both the start and the goal state at the same time, as it is shown in Fig. 3.12. The bidirectional reasoning procedure terminates when the reasoning "bridge" seen in the Figure is built up.

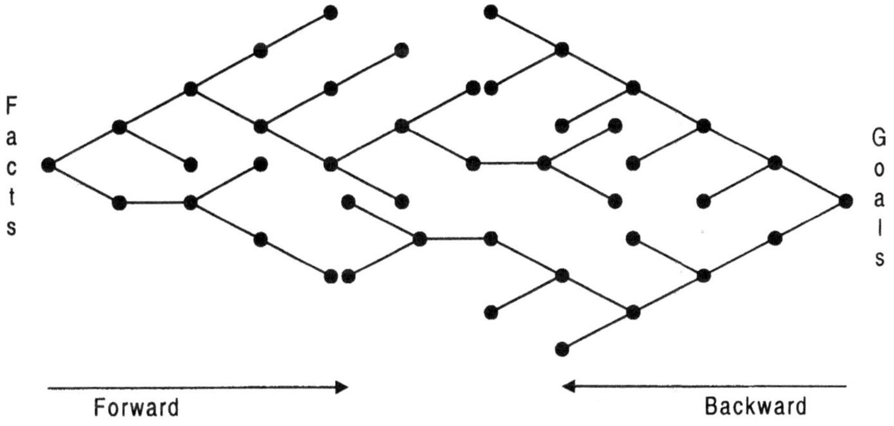


*Figure 3.12.* Bidirectional reasoning

# 5. SEARCH METHODS

As it was mentioned earlier, reasoning problems are solved by *search* on the reasoning graph in the state-space. Search in itself is a general problem solving method or mechanism. Search is used in order to get from the initial state to one or more possible goal states during problem solving. The solution is described by a *path*, which consists of rules or transitions executed one after the other, starting at the initial state and ending in the goal state.

We have also seen that the inference engine often gets to a decision position during reasoning or search when it applies conflict resolution techniques. A *search strategy* is used during search for decision making. It is often supported by concrete knowledge about the task to be solved, called *heuristics*.

We can group search strategies into two main categories:

- *non-modifiable control strategies*

  Non-modifiable control strategies attempt to get from the initial state to a goal state supposing that all of the chosen rules have been selected properly. There is no opportunity to withdraw the application of a rule, to modify the strategy or to try the other applicable rules during the search.

- *modifiable control strategies*

  Modifiable control strategies are able to recognize the erroneous or improper application of a rule. It may happen during the search that we reach a stage which does not lead to a goal state or where it does not seem promising to resume the search in that direction. In such a state the algorithm backtracks to an earlier state and a new direction is chosen in order to find the goal state.

Search strategies can be divided into two groups from the viewpoint of the application of heuristics:

- *uninformed control strategies*

  In an uninformed control strategy, all of the paths are traversed in a systematic way. There is no information about the "goodness" of the path or a node examined in a nongoal state. The algorithm can only distinguish a goal state from a nongoal state. An uninformed search strategy is also called *blind search strategy.*

- *informed control strategies*

  Here the specific knowledge about the given problem is also used. The informed control strategy is called *heuristic control strategy* or *heuristic search.*

The general and some important special search methods are introduced and discussed in the following sections.

## 5.1 THE GENERAL SEARCH ALGORITHM

This section describes a general algorithm that searches for a solution path in a graph. The essence of the method is to register all of the examined paths that started from the initial state. The method makes it possible to move along the path which promises to be the best from the aspect of reaching the goal node. Then all the successors of the node in the starting point of the selected path are produced. This is called the *expansion* of the node, whereby a subgraph of the representation graph is constructed. The expansion of the graph is finished if a goal node is reached.

The main steps of the general search algorithm are as follows:

1. Add the initial node representing the initial element to $L$ representing the list of nodes that have not yet been examined.

2. If $L$ is empty, fail. Otherwise, choose a node $n$ from $L$.

3. If $n$ is a goal node stop and return it and the path from the initial node to $n$.

4. Otherwise, remove $n$ from $L$, expand the nodes of $n$ (produce the subsequent nodes to $n$) and add them to $L$. Return to step 2.

$L$ is called the *list of open nodes* (the nodes which are expanded but not examined). The methods of selection from this list define different search algorithms. In practice the values of a function (the so called evaluation function) are often used for choosing an open node from a list.

## 5.2 DEPTH-FIRST SEARCH

*Depth-first search* is one of the uninformed strategies. The simplest way to understand how depth-first search expands the nodes of the search tree is to look at Fig. 3.13. The numbers appearing as labels at the nodes of the tree show the order the nodes are examined by the depth-first search algorithm.

It is always one of the nodes at the deepest level of the tree that is expanded (nodes are examined from left to right). When a terminal node (with no expansion) but not a goal node is reached, the procedure backtracks and expands nodes at shallower levels.

Depth-first search can be implemented by pushing the children of a given node into the *front* of list $L$ in step 4. of procedure in section 5.1 of this Chapter and always choosing the *first* node from $L$. The open list is used as a *stack*.

The advantages of the method are its easy implementation and modest memory requirement. The drawbacks of depth-first search are that it can get stuck in an infinite loop and never return a solution, and it can find a solution that is longer (or more expensive) than the optimal solution. So depth-first search is neither complete nor optimal.

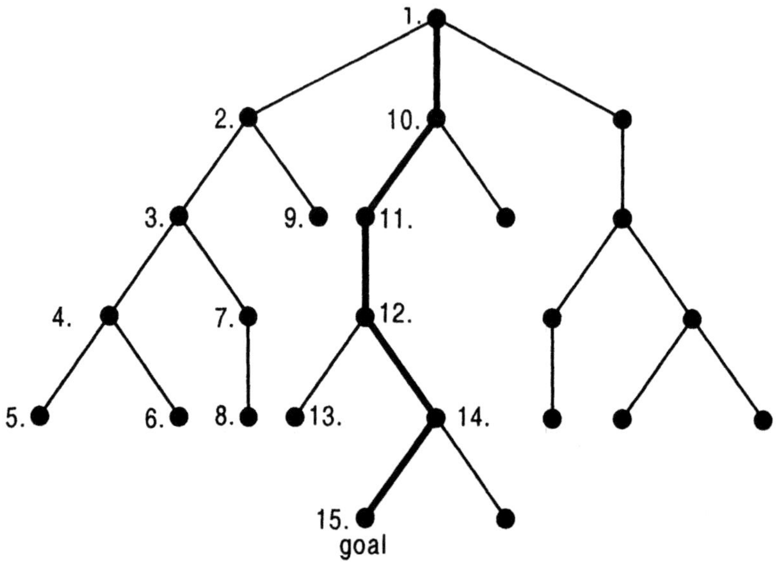


*Figure 3.13.* **The depth-first search**

## 5.3 BREADTH-FIRST SEARCH

The other uninformed strategy, ***breadth-first search*** avoids the drawbacks of depth-first search. As Fig. 3.14 shows, the breadth-first search algorithm examines the nodes at a certain depth only if all the nodes at shallower depths have been examined.

Breadth-first search can be implemented by pushing the children nodes of a given node into the *back* of list $L$ in step 4. of procedure in section 5.1 of this Chapter and always choosing the *first* node from $L$. The open list is used as a *queue*.

The advantage of breadth-first search is that it always finds a solution if it exists and the solution is always optimal. The drawback of the method is that its memory requirement increases exponentially with the size of the problem.

The method of search is often determined by the knowledge of problem structure. For example, depth-first search is used when there are only a few consequences of a state that have long reasoning chains, and breadth-first search is used when there are many consequences with short reasoning chains.

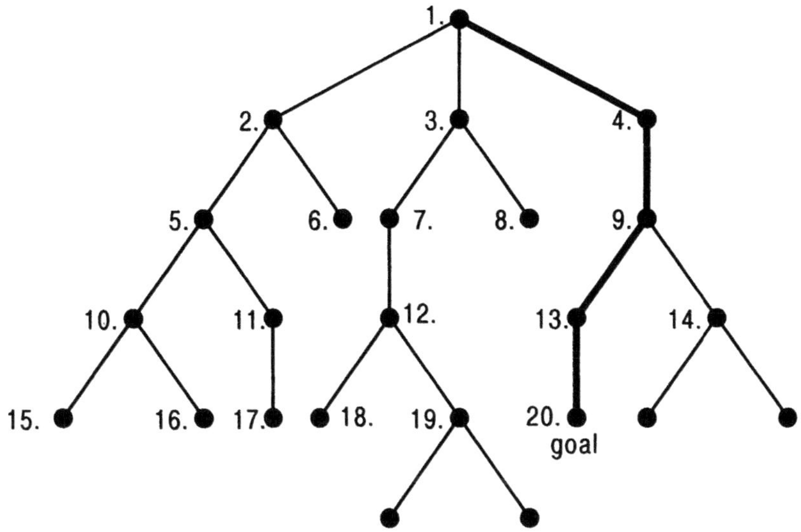


*Figure 3.14.* The breadth-first search

## 5.4 HILL CLIMBING SEARCH

*Hill climbing search* is the most known non-modifiable search strategy. An appropriate heuristic function, which takes its minimal value in the initial node and its maximal value in the goal node is used for choosing the next node. The problem is solved by a special maximum search in the state-space. As can be seen in Fig. 3.15, the algorithm examines all the successors of the current node, selects the successor with the highest heuristic value, uses that as the next node to search from and stops when no successors has a higher value than the current node. The method is known as gradient method beyond AI. Of course, the hill climbing method is suitable finding the minimum value, too.

Some important difficulties can occur during hill climbing search, which are as follows:

- *local maxima*: the search has found a local maximum, but has not found the global maximum

- *plateaus*: the search has reached a node, and around it the evaluation function is essentially flat

- *ridges*: the search has reached a node where the values of the successors are lower, but a node with higher value can only be reached by the combination of several steps

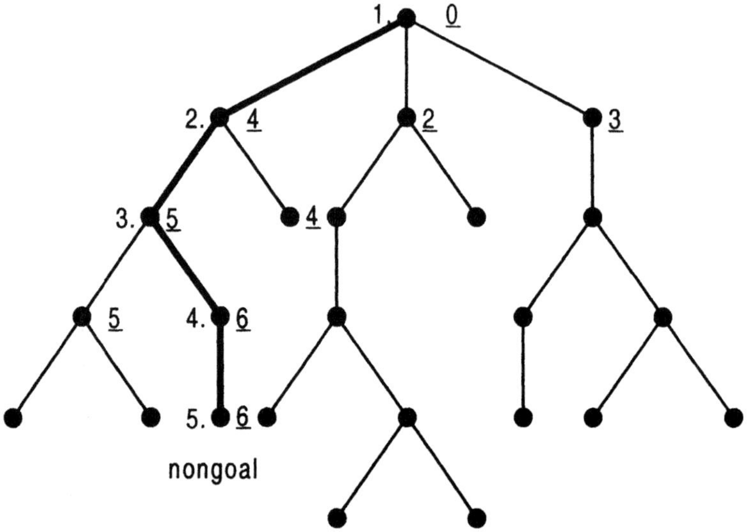


*Figure 3.15.* The hill climbing search (the value of the heuristic function is denoted by the underlined numbers)

The advantage of hill climbing search is its small memory requirement. Moreover, if the algorithm is started from a good starting point then the goal is reached quickly.

## 5.5 A* SEARCH

*A* search* is a well-known and efficient heuristic search method. In this method a heuristic function $f(n)$ is used to estimate the cost of the cheapest solution through the node $n$.

$$f(n) = g(n) + h(n)$$

The heuristic function is the sum of the cost of the path from the initial node to the current node denoted by $g(n)$ and the estimated cost from the current node to the goal denoted by $h(n)$:

$$\underbrace{\underbrace{start \overset{actual}{\rightarrow} n}_{g(n)} \underbrace{\overset{estimate}{\rightarrow} goal}_{h(n)}}_{f(n)}$$

As Fig. 3.16 shows A* search always expands one of the nodes with the lowest cost. It can be implemented by *ordering the open nodes in list L according to* $f(n)$ and always choosing the node with the *lowest cost* in $L$ in step 2. of procedure in section 5.1 of this Chapter.

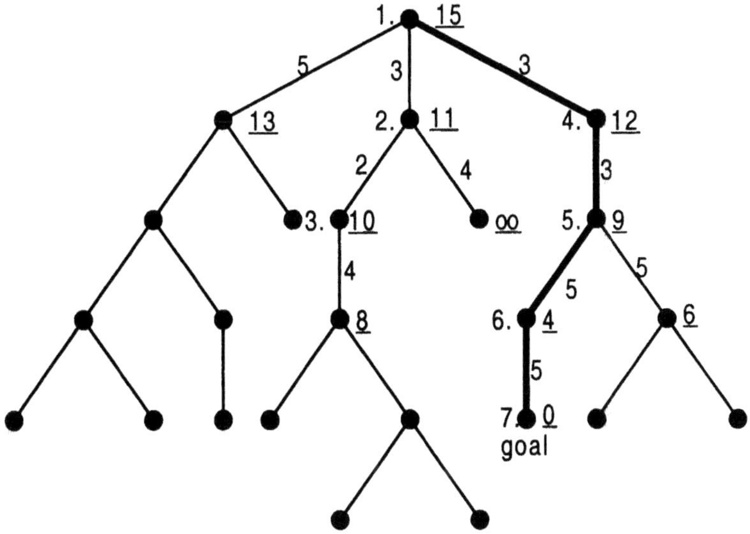


*Figure 3.16.* The A* search (the value of the heuristic function is denoted by the underlined numbers and the numbers labeling the arcs denote their costs)

If the $h(n)$ function used by the algorithm is constructed in such a way that it never *overestimates* the cost to reach the goal, then it is guaranteed to find the optimal solution. Such a $h(n)$ is called an *admissible heuristic*. If the value of the function $h(n)$ is equal to zero for every node and there are unit costs of arcs, then the A* search reduces to the breadth-first search.

Chapter 4

# VERIFICATION AND VALIDATION OF RULE-BASED KNOWLEDGE BASES

Knowledge representation tools and techniques are able to store and handle quite complex knowledge bases with a high number of complicated relations over a massive set of facts. As we have already seen in Chapter 2 the dominance of complex relations characterizes knowledge bases in comparison with traditional databases. Therefore, it is extremely important to construct and maintain knowledge bases with high quality, that is with reliable and solid content. The procedures for verification and validation of knowledge bases are therefore of primary importance [22], [23], [24], [25].

We can test a knowledge base in two principally different ways.

- Either we *validate* it by comparing its content with additional knowledge of a different type [26],
- or we *verify* it by checking the knowledge elements against each other to find conflicting or missing items.

Because of the great variety and flexibility of knowledge representation tools and techniques, it is almost impossible to give a general approach of verification and validation of knowledge bases. Therefore we shall restrict ourselves to the simplest case when *the knowledge base only contains rules in datalog format* [27]. Such knowledge bases will be called *rule-based knowledge bases* or shortly *rule-bases.*

It is important to note, however, that we may have hidden rules to a datalog rule-base which describe semantical relationships between predicates and these rules may contain negation as well. Such rules naturally arise when a natural rule-base is transformed to its datalog format (see in subsection 2.3 of Chapter 2). The hidden rules destroy the datalog property of the rule-base when they are taken into account during verification.

The verification of completeness and contradiction freeness of rule-based knowledge bases is described and analyzed in this chapter using the notions and techniques of theoretical computer science [28].

We shall consider the following important verification properties separately in the following sections:

- contradiction freeness
- completeness

In both cases, the notion of the property is followed by the description of its verification procedure as a standard algorithmic decision problem.

It is important to note that the abstract data structure (3.1) introduced in Chapter 3 will be used here to describe the structure of a datalog rule set:

$$\begin{array}{rcl} \mathit{rule\_base} & = & \{ \ (n_P \ , \ n_R) \\ (r_1) & : & (p_{1_1} \wedge \cdots \wedge p_{1_{n_1}}) \ \rightarrow \ q_1; \\ \ldots & & \ldots\ldots\ldots\ldots\ldots \\ (r_{n_R}) & : & (p_{n_{R1}} \wedge \cdots \wedge p_{n_{Rn_R}}) \ \rightarrow \ q_{n_R}; \\ & & \} \end{array} \tag{4.1}$$

where $n_P$ is the number of predicates and $n_R$ is the number of rules.

# 1. CONTRADICTION FREENESS

One of the most important requirements for knowledge bases is that their content should not have any contradiction neither formal (syntactical) nor semantical. Syntactical or formal contradictions are investigated by the verification process of the knowledge base that examines contradiction freeness.

## 1.1 THE NOTION OF CONTRADICTION FREENESS

Reliable knowledge bases have a unique primary or inferred knowledge item, if they have any, irrespectively of the way of reasoning. This property is described in precise mathematical terms by the notion of contradiction freeness for rule-based knowledge bases.

**Definition 4.1.** *A rule-based knowledge base with a data structure (4.1) is contradiction free if the value of any of the non-root predicates is* *uniquely determined by the rule-base using the rules for forward chain reasoning.*

## 1.2 TESTING CONTRADICTION FREENESS

In order to analyze how one can test contradiction freeness of a rule-base in datalog format, we formulate testing as a standard algorithmic decision problem as follows.

TESTING CONTRADICTION FREENESS

*Given:*

- *A rule-based knowledge base* with its abstract data structure (4.1)

*Question:*
Is the rule-base contradiction free?

*Solution:*
From the definition above it follows that we need to compute the value of each non-root predicate under every possible circumstance, that is with every possible set of the root predicate values and in every possible way. Therefore, the following substeps should be performed to check the contradiction freeness of the given rule-base.

1. *Determine the set of root predicates*
   by analyzing the dependence graph of the datalog rule set or by collecting all predicates which do not appear on the consequence part of any rule. This is a polynomial step.

2. *Construct the set of all possible values for the root predicates* (to be stored in the set $S_{rp}$)
   Here we have to consider all the three possible values **true**, **false** and **unknown** for every root predicate. From the viewpoint of reasoning, however, the values **false** and **unknown** are equivalent, therefore the number of the elements in this set is $2^{n_{rp}}$. This implies that this step is not polynomial.

3. *For every element in $S_{rp}$ perform forward chaining and compute the value of the non-root predicates in every possible way*
   that is by applying the rules in every possible order. This step requires to solve a FORWARD CHAINING search problem (see section 2. of Chapter 3) for every possible value of the root predicates. Therefore this step is usually *NP-complete* .

4. Finally, *check that the computed values for each of the non-root predicates are the same.* If yes then the answer to our original question is *yes*, otherwise *no*.

It is important to note that we only check whether we have a unique computed value of every predicate if there exists any. It means that we do not require that the value of every predicate is determined from every given set of root predicates by the forward chaining.

It is worth noting that there is a strong procedure type relationship between TESTING CONTRADICTION FREENESS mentioned above and FORWARD CHAINING problems because the former calls the latter as a procedure in step 3.

The following simple example illustrates the notion of contradiction freeness.

---

**EXAMPLE 4.1** A simple rule set with contradiction

Consider a simple rule set defined on the following set of predicates:

$$P = \{p_1, p_2, p_3, p_4, p_5\} \tag{4.2}$$

so that $p_5 = \neg p_4$ holds. This relationship is described by a "virtual" rule pair:

$$\begin{array}{rcl} (r_{01}) & : & p_5 \rightarrow \neg p_4; \\ (r_{02}) & : & p_4 \rightarrow \neg p_5; \end{array}$$

Let the implication form of the rule set be

$$\begin{array}{rcl} (r_1) & : & (p_1 \wedge p_2) \rightarrow p_4; \\ (r_2) & : & (p_3 \wedge p_1) \rightarrow p_5; \\ (r_3) & : & (p_1 \wedge p_2) \rightarrow p_3; \end{array}$$

Then the number of predicates $n_P = 5$ and the number of datalog rules $n_R = 3$ can easily be computed as well as the set of root predicates

$$P_{root} = \{p_1, p_2\}$$

Let us have the following values for the root predicates:

$$p_1 = \textbf{true} \quad , \quad p_2 = \textbf{true}$$

Then we get for $p_4$ the following values

- **true** from $(r_1)$
- **false** from $(r_3)$, $(r_2)$, $(r_{01})$

Observe that the contradiction is caused by the presence of the hidden rules in the rule set.

---

## 1.3 THE SEARCH PROBLEM OF CONTRADICTION FREENESS

The verification of a rule-based knowledge base can be performed in two principally different ways depending on the strategy the knowledge base is constructed.

- *global verification*
  Here the whole rule-based knowledge base is constructed first and the verification is performed thereafter in one shot. Then the solution of the decision problem TESTING CONTRADICTION FREENESS gives only a *"yes/no"* answer with no indication on where and how the contradiction may arise.
- *incremental verification*
  The other way to build a knowledge base is to extend it incrementally, that is to add a single (or a few) new rules to an already verified rule-base. Then verification is also performed in each extension step and it is clear that the possible problems are related to the new part.

In both cases the source of the possible contradiction problems can be found by analyzing the way contradicting value(s) have been generated for some of the non-root predicates. This requires the generation and analysis of the whole set of reasoning trees obtained during the solution of the decision problem TESTING CONTRADICTION FREENESS. This can be done if the search equivalent of this problem is solved. It is in the following form.

ANALYZING CONTRADICTION FREENESS

*Given:*

- *A rule-based knowledge base* with its abstract data structure (4.1)

*Compute:*
the whole set of possible reasoning trees to generate all possible values of the non-root predicates.

*Solution:*
By comparing the problem statement above to that of TESTING CONTRADICTION FREENESS it can be seen that the ANALYZING CONTRADICTION FREENESS problem is *NP*-hard both from the viewpoint of time and space.

# 2. COMPLETENESS

Completeness is a dual problem of contradiction freeness in a certain sense because here one is interested in whether the knowledge in the knowledge base is enough to solve the given problem.

## 2.1 THE NOTION OF COMPLETENESS

Rich enough knowledge bases have an answer (even this answer is not unique) to every possible query or question. This property is formulated in a rigorous way by the notion of completeness in case of rule-based knowledge bases.

**Definition 4.2.** *A rule-based knowledge base with a data structure (4.1) is complete if any non-root predicate gets a value when performing forward chain reasoning with the rules.*

## 2.2 TESTING COMPLETENESS

Similarly to the case of testing contradiction freeness, we formulate testing completeness as a standard algorithmic decision problem as follows.

TESTING COMPLETENESS

*Given:*

- *A rule-based knowledge base* with its abstract data structure (4.1)

*Question:*
Is the rule-base complete?

*Solution:*
From the definition it is seen that now we do not need to compute the value of each of the predicates in every possible way but we need to find out if every non-root predicate is present in the reasoning tree in all cases. Therefore, completeness can be tested by the following steps.

1. *Determine the set of root predicates*
   by analyzing the dependence graph of the datalog rule set, for example. This is a polynomial step.

2. *Construct the set of all possible values for the root predicates* (to be stored in the set $S_{rp}$)
   The number of the elements in this set is $2^{n_{rp}}$, therefore, this step is not polynomial.

3. *For every element in $S_{rp}$ perform forward chaining and generate a reasoning tree*
   until either all non-root predicates appear at least once or all the rules have been applied in every possible order. This step requires the solution of a FORWARD CHAINING search problem (see section 2. in Chapter 3) for every possible value of the root predicates. Therefore, this step is usually *NP-complete.*

4. Finally, *check that each of the non-root predicates gets at least one value in every possible case.* If yes, then the answer to our original question is *yes*, otherwise *no.*

A simple example of a non-complete rule set, which is exactly the same as in Example 4.1, is given below.

---

**EXAMPLE 4.2** A simple non-complete rule set

Consider a simple rule set defined on the same set of predicates (4.2) as in Example 4.1. The "virtual" rule pair ($r_{01}$) and ($r_{02}$) is also associated with the set of predicates.

Let the implication form of the datalog rule set be the same as the rules ($r_1$)-($r_3$).

Let us have the following values for the root predicates:

$$p_1 = \textbf{true} \quad , \quad p_2 = \textbf{false}$$

Then we have no applicable rule from the rule set therefore the non-root predicates $p_3$, $p_4$ and $p_5$ are undetermined in this case.

---

## 2.3 THE SEARCH PROBLEM OF COMPLETENESS

The need to formulate and solve the search problem related to TESTING COMPLETENESS arises the same way as it is explained in the subsection 1.3 that describes the search problem of contradiction freeness.

This problem formulation and solution technique is used if one wants to obtain information on how the non-completeness problem(s) arise.

ANALYZING COMPLETENESS

*Given:*

- *A rule-based knowledge base* with its abstract data structure (4.1)

*Compute:*
the whole set of possible reasoning trees to generate all possible values of the non-root predicates.

*Solution:*
By comparing the problem statement above to that of TESTING COMPLETENESS it can be seen that the ANALYZING COMPLETENESS problem is *NP-hard* both from the viewpoint of time and space.

# 3. FURTHER PROBLEMS

This section contains important extensions and consequences of the contradiction freeness and completeness sections before.

## 3.1 JOINT CONTRADICTION FREENESS AND COMPLETENESS

In practice, one needs knowledge bases which are *both* contradiction free and complete. If one compares the principal steps of the two testing algorithms we can observe that generating steps 1.-3. are exactly the same, it is only the evaluation of the generated reasoning tree that is different. This calls for the combination of the two algorithms, that is checking contradiction freeness and completeness by one single algorithm that consists of the joint steps 1.- 3. and of the combined evaluation steps 4.

Because of the *NP-hard* computational complexity of the test of contradiction freeness and completeness, approximate procedures have also been proposed [29].

## 3.2 CONTRADICTION FREENESS AND COMPLETENESS IN OTHER TYPES OF KNOWLEDGE BASES

The notion of and testing procedures for contradiction freeness and completeness have been introduced and discussed only for the most simple case, that is for knowledge bases only consisting of datalog rules possibly extended by hidden rules.

There are a number of issues which make it difficult to generalize the notions and algorithms to other types of knowledge bases.

1. *Knowledge items with non-Boolean or non-deterministic values*
   The presence of non-Boolean and/or uncertain values in the knowledge base makes it difficult to compare the value of the non-root predicates (or knowledge items) obtained by different ways of reasoning. This calls for an extension of the definitions of contradiction freeness and completeness.

   In this case, one should use suitably defined ***knowledge comparison norms***, similarly to the case when vectors or matrices are compared. More about this problem can be found in Chapter 9, which deals with completeness and contradiction freeness of fuzzy rule-bases when uncertainty is present.

2. *Special non-rule-based reasoning methods*
   If the knowledge base contains other knowledge elements than predicates and rules, then usually special reasoning methods need to be applied to obtain causes or consequences of a given knowledge set.

   In this case, not only the definitions of contradiction freeness and completeness should be extended but the conceptual steps of the solution of both the corresponding decision and search problems should also be completely changed.

# 4. DECOMPOSITION OF KNOWLEDGE BASES

The *NP-hard*ness of both the testing of contradiction freeness and completeness even in the simplest case of rule-based knowledge bases requires an attempt to constrain the size of the knowledge base part to verify, that is both the number of predicates and the number of rules [30]. This can be done by decomposing the knowledge base into parts which are internally strongly dependent but "loosely dependent" on the knowledge belonging to other parts.

This way one can create a ***hierarchical decomposition*** structure of a rule-base by partitioning the predicates into classes and associating the rules which only depend on predicates of a given class to that class. The rules with predicates in more than one class become member of the higher, inter-class knowledge representation level.

The problems and challenges of decomposing knowledge bases are explained here using the knowledge bases of the most simple structure as an example: rule-based knowledge bases. Decomposition techniques use graphs to represent the structure of a datalog rule-base: the dependence graph of the datalog rule set (see in section 2.3 of Chapter 2).

## 4.1 STRICT DECOMPOSITION

The strict decomposition of a rule-based knowledge base is carried out by computing the *strong components of the dependence graph.* We recall that a strong component of a directed graph is a set of vertices such that any (ordered) pair of vertices from the set is connected by a directed path. The predicates belonging to a strong component together with the rules forming the directed edges within the set (that is the induced subgraph generated by the strong component) form one class. Next, all the inter-class rules will form a hyper-graph of no loops. *The decomposition of the dependence graph into strong components is a polynomial step, therefore the strict decomposition is also polynomial.*

Unfortunately, the whole rule-base may easily form one single strong component in most cases that are useful from the practical point of view.

## 4.2 HEURISTIC DECOMPOSITION

Heuristic decomposition is needed when the dependence graph forms one single strong component due to the strong inter-relationships between the predicates. Here heuristic considerations as well as semantic arguments on the meaning of the predicates and rules can be and should be used to obtain a "good enough" decomposition.

The goal of decomposition is to form sub-graphs within the dependence graph such that

- the size of the sub-graphs both in the number of its vertices and in the number of its induced edges are below a limit,
- the vertices of the sub-graphs form a partition in the vertex set of the overall graph,
- there are "as few as possible" edges between the sub-graphs.

It is easily seen that the optimal version of the above problem leads to a GRAPH ISOMORPHISM problem which is known to be *NP*-hard. Therefore, the exact solution is not feasible, heuristic methods should be applied.

Chapter 5

# TOOLS FOR KNOWLEDGE REPRESENTATION AND REASONING

This chapter introduces and compares the most important traditional tools for knowledge representation and reasoning. Of course, there is a wide selection of tools available from which we had to choose. Because of their theoretical and practical value and popularity, the following tools have been selected:

- Lisp programming language [31] - [35]

- Prolog programming language [36] - [40]

- Expert system shells [41] - [45]

The tools are arranged and introduced in the order of their level of conceptual complexity.

Lisp can be regarded as a general purpose assembly level language, which is almost only based on the notion of and operations on lists.

Prolog is a high-level declarative language and reasoning environment with a built-in inference engine.

Finally, expert system shells are the most sophisticated environments for prototyping and implementing an expert system.

When describing the various knowledge representation and reasoning tools, we use a number of program parts for illustration purposes. The string the user enters and the answer that is given are distinguished by `teletype font` typesetting.

# 1. THE LISP PROGRAMMING LANGUAGE

Lisp is a functional programming language that takes its name from List Processing. It is used for manipulating on symbols. It evaluates procedures using the notion of a mathematical function.

Lisp was developed in the late 50s by John McCarthy in the USA. There are several Lisp dialects but all of them kept the fundamental elements of the first version. Later on Common Lisp has become popular and is now extensively used because it is widely available and is an accepted standard for commercial use.

In Lisp programs all of the problems can be described in the form of function calls. Some important characteristics of the language are:

- the construction of programs and data is the same,
- Lisp programs can produce and can execute other programs, and
- they can even modify themselves.

## 1.1 THE FUNDAMENTAL DATA TYPES IN LISP

The basic elements like `5, a23, +, 2.5, T, NIL` are word-like objects called *atoms* in Lisp. The atoms consist of any number of digits and characters. There are two types of atoms: *numeric atoms* or *numbers* like `5, 2.5` and *symbolic atoms* or *symbols* like `a23, +, T, NIL`. `T` and `NIL` are special symbols for the logical true and false values.

We can build sentences in the form of *lists*, for example `(a b c)`, `(x 1)`, `((a) (3 4))`, `()`. Lists consist of a left parenthesis, zero or more atoms or lists separated by a space and a right parenthesis. As you can see, the definition of the list is recursive, the elements of a list can also be lists of any depth. A list containing no elements is called an *empty list* and is denoted by `()` or `NIL`. Procedures, procedure call statements and data are all stored in lists. The atoms and lists together are called *symbolic expressions* or *expressions*. This way both programs and databases consist of expressions.

Fig. 5.1 depicts the hierarchy of basic data types in Lisp.

Let us now examine the properties of a list in detail. The first element of a list is the *head* and the rest is the *tail*. A *tail* may be composite, that is it may contain several elements.

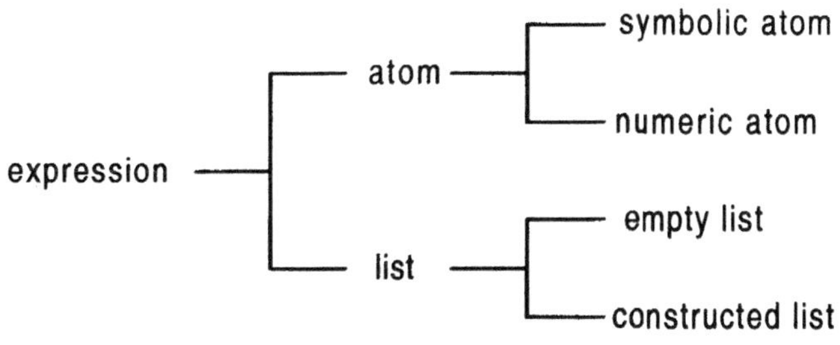


*Figure 5.1.* The basic data types in Lisp

$$\underbrace{(element_1}_{head} \quad \underbrace{element_2 \quad \ldots \quad element_n)}_{tail}$$

In a list describing a procedure in a Lisp program, the head is a *procedure name* and the tail contains the *arguments* the procedure works with. This so-called *prefix notation* makes the unification of all procedure declaration and call possible, because the procedure name is always in the same place, no matter how many arguments are involved.

Syntactically, a list can be imagined as a tree. The root of the tree is the list being examined, the leaf nodes are the atoms and the other nodes are the elements of the list. The depth of the tree is equal to the depth of the list, so the first level of the tree corresponds to the top-level elements in the list.

The following simple example illustrates the concept of multi-level lists.

---

**EXAMPLE 5.1** A simple list with its syntax tree

Consider the following simple list:

(+ (∗ 2 3) (−4 1))

with depth 2. The syntax tree of this list is shown in Fig. 5.2.

---

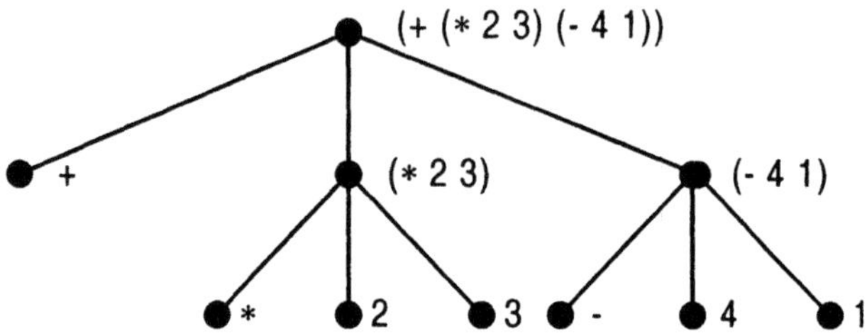


*Figure 5.2.* The representation of a list with a graph

## 1.2 EXPRESSIONS AND THEIR EVALUATION

There are several expressions in a Lisp program used to solve a problem. Their evaluation and role in the program can be different.

Lists of the first type describe procedures. The Lisp program is executed by calling these procedures. Remember that a procedure call is also in the form of a list, where the head of the list is the procedure name and the rest of the elements are the arguments in the following general form:

$$(< procedure\ name > < argument_1 > \ \ldots \ < argument_n >)$$

The number of arguments depends on the type of the procedure. There are procedures (for example `+`, `LIST`, etc.) where the number of arguments may vary.

Users can even define such procedures. The procedures supplied by Lisp itself are called *primitives* and the procedures created by the user are called *user-defined procedures.*

Every expression (atom and list) has a value and the Lisp interpreter reads, evaluates and prints these values in an endless cycle. When you start a Lisp system it displays a *prompt* to tell you that it is waiting for the input data. In Common Lisp the prompt is an asterisk:

```
*
```

You can type the input and observe the output.

```
* (+ (* 2 3) (- 4 1))
9
```

The response of Lisp is the value of the expression printed after the asterisk, which in this case is 9. The arguments of the expression can be procedures and their arguments and even the head of the procedure can be another procedure. The algorithm is evaluated as follows:

1. evaluation of the head (it must be a predefined procedure name)
2. evaluation of the first, second, ... argument (the second, third, ... element of the list)
3. using the procedure (the value of the head) with the arguments.

As in other programming languages, there are variables in Lisp, too. Variables don't have to be declared in Lisp. Symbols are used for storing values. The value of a number is the proper number and the value of a symbol is not bound at first. Values can be set in different ways, for example with the `SETF` primitive discussed in section 1.3.3 in this Chapter. There are no variable types in Lisp, so the value of a symbol is optional.

## 1.3 SOME USEFUL LISP PRIMITIVES

There are several Lisp primitives used to set values, use lists and arithmetic expressions, organize cycles, handle files, write procedures etc. In this section some of the most frequently used primitives are introduced and discussed.

### 1.3.1 THE QUOTE PRIMITIVE

It was mentioned earlier that the syntax of programs and data is the same. The interpreter cannot distinguish between them so it needs help from the user. The `QUOTE` primitive is used for differentiating between program and data. It stops the evaluation procedure and a quoted expression can be used as data.

```
* (quote (+ 1 6))
(+ 1 6)
```

without quote:

```
* (+ 1 6)
7
```

`QUOTE` is a frequently used primitive and ' is a short notation equivalent to it.

```
* '(+ 1 6)
(+ 1 6)
```

As you can see, the same expression can be data at one time and a program at another. An expression is considered to be data when it is not evaluated, and it is a program part when it is evaluated. In the Lisp language, the interpretation of an expression is dynamically assigned to the expression during evaluation.

### 1.3.2 PRIMITIVES MANIPULATE ON LISTS

Since there are several list expressions in a Lisp program, it is important to know primitives that manipulate lists. First the basic primitives for dissecting lists are described.

The FIRST (or in old programs CAR) primitive selects the first top-level element from its list argument.

```
* (first '(x y z))
X

* (car '((1 2) (a b)))
(1 2)
```

The REST (or CDR) primitive performs a complementary operation: it returns a list that contains all but the first top-level element.

```
* (rest '(x y z))
(Y Z)

* (cdr '((1 2) (a b)))
((A B))
```

It is important to remember that REST always returns a list. When REST is applied to a list with only one or zero element it returns the empty list and when FIRST is applied to the empty list the result is the empty list by convention.

```
* (rest '(a))
NIL

* (rest ())
NIL

* (first ())
NIL
```

Several composite primitives can be constructed from CAR and CDR in the form of CXXR, CXXXR, CXXXXR, where X denotes either an A denoting CAR or a D denoting CDR. With this convention the following expressions are the same:

```
(cdar '((1 2) (a b))) ≡ (cdr (car '((1 2) (a b))))
```

Of course, the evaluation of such an expression starts with the inner list, so the value of the expression is the following:

```
* (cdar '((1 2) (a b)))
(2)
```

Another group of primitives is used for constructing lists.

The CONS primitive attaches the expression given as its first argument at the front of the list given in its second argument.

```
* (cons 'x '(y z))
(X Y Z)

* (cons '(a b) '(c d))
((A B) C D)
```

The parts of a list decomposed by the FIRST and REST primitives can be used for reconstructing the original list by CONS as it is shown in Fig. 5.3.

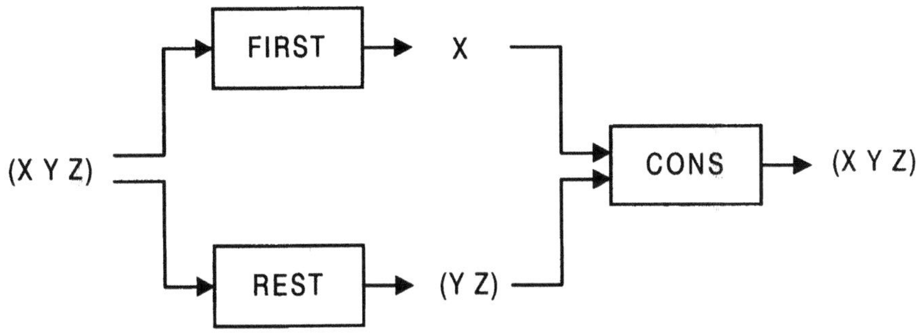


*Figure 5.3.* The relationship between FIRST, REST and CONS

APPEND concatenates the top-level elements of the lists in its arguments into a single list.

```
* (append 'x '(y z))
ERROR (about that the arguments must be lists)

* (append '(a b) '(c d))
(A B C D)
```

The LIST primitive constructs a list from the expressions in its arguments.

```
* (list 'x '(y z))
(X (Y Z))

* (list '(a b) '(c d))
((A B) (C D))
```

LIST and APPEND work on any number of arguments, that is on more than two arguments.

```
* (list (+ 1 2) (* 3 4) '(a b))
(3 12 (A B))
```

```
* (append '(1 2) '((3 4)) '(a b))
(1 2 (3 4) A B)
```

### 1.3.3 ASSIGNMENT PRIMITIVES

In Lisp, symbols may have values associated with them. The special symbols and numbers always have values, this value is the symbol itself and it cannot be changed. Programmers can assign values to other symbols with the help of the SETF or SET primitive.

```
* (setf ab-list '(a b))
(A B)
```

The SETF primitive evaluates its second argument and stores the resulting value in memory assigned to the first argument, which should be a symbol identifier. SETF is not a usual procedure, because it does not evaluate its first argument and it does more than just returning a value: it assigns the value of the second argument to the symbol in the first argument.

```
* ab-list
(A B)
```

The SETF primitive can handle more symbol-value pairs. Then the values of the even arguments are assigned to the arguments before.

```
* (setf ab-list '(a b) xy-list '(x y))
(X Y)
```

The return value is then the value of the last argument.

The SET primitive works like SETF, but it evaluates its odd arguments, too.

```
* (set (first '(a b c)) 123)
123
* a
123
```

### 1.3.4 ARITHMETIC PRIMITIVES

In Lisp all the standard arithmetic functions are available. These are: `+, -, *, /, mod, sin, cos, tan, sqrt, expt, min, max,` etc.

All of them accept any kind of number (integer, real, rational, complex) as an argument and the type of the return value depends on the types of the arguments. Some examples below illustrate the properties and use of arithmetic primitives:

```
* (/ 1.5 0.6)
```

```
2.5

* (/ 9 3 3)
1

* (/ 7 3)
7/3

* (sqrt -9)
#C(0.0 3.0)

* (min (+ 1 1) (* 2 2) 3)
2
```

### 1.3.5 PREDICATES

The procedure that returns a true or false logical value is called *predicate.* For the notation of the false value, the special symbol `NIL` is always used and the true value is often denoted by the special symbol `T`. In general, anything other than `NIL` denotes a logical true value.

One group of predicates examines the equality of two expressions. For example, numerical equality is determined by the = predicate, the equality of symbols is determined by the `EQ` predicate and the equality of expressions is determined by the `EQUAL` predicate. The following simple examples illustrates the use of the primitives above:

```
* (= (+ 1 2) 3.0)
T

* (= 'a 5)
ERROR (about that "a" is not a number)

* (eq 'b (first '(b c)))
T

* (equal (+ 2 2) 4)
T

* (equal 'a 5)
NIL

* (equal (list 'a (first '(2 3))) '(a 2))
T
```

The `MEMBER` predicate tests whether its first argument is a top-level element of the list in its second argument. If the first argument is not found in the list, `NIL` is returned, otherwise the tail of the list beginning with the first argument is returned, as it can be seen in the examples below.

```
* (member 'element '(the element is in the list))
(ELEMENT IS IN THE LIST)

* (member 'element '(not in the list))
NIL

* (member 'element '((not top-level element)))
NIL
```

Lisp has several primitives that test whether an expression corresponds to a particular data type. The ATOM predicate tests its argument to see if it is an atom, NUMBERP examines if it is a number, SYMBOLP tests for a symbol and LISTP for a list.

```
* (atom (first '(1 2 3)))
T

* (atom (rest '(1 2 3)))
NIL

* (numberp (first '(1 2 3)))
T

* (numberp (rest '(1 2 3)))
NIL

* (symbolp (first '(1 2 3)))
NIL

* (symbolp (first '(a b c)))
T

* (listp (first '(1 2 3)))
NIL

* (listp (rest '(1 2 3)))
T
```

There are two predicates that check whether the argument is an empty list: NULL and ENDP. The difference between the two predicates lies in the type of argument: the argument type of the NULL predicate is optional but in ENDP the argument must be a list.

```
* (null (first '(a)))
NIL

* (null (rest '(a)))
T

* (endp (first '(a)))
ERROR (about that argument must be a list)
```

```
* (endp (rest '(a)))
T
```

Lisp provides three logical predicates: AND, OR, and NOT. AND and OR can have any number of arguments, which are evaluated from left to right. AND returns NIL if any of its arguments evaluates to NIL and none of the remaining arguments is evaluated. In all other cases, it returns the value of the last argument. OR returns NIL if all of its arguments evaluate to NIL, otherwise it returns the value of the first non-NIL argument and the remaining arguments are not evaluated. The NOT predicate alters the truth value of its argument: it turns a non-NIL value to NIL and NIL to T. Simple examples are:

```
* (and (setf x 3) (member 'b '(a b c)))
(B C)
* x
3

* (and (numberp 'a) (setf y 12))
NIL
* y
ERROR (about that "y" has not bounded)

* (or (member 'b '(a b c)) (setf y 12))
(B C)
* y
ERROR (about that "y" has not bounded)

* (or (numberp 'a) (null '(1 2 3)))
NIL

* (not 'a)
NIL

* (not (member 'x '(a b c)))
T
```

### 1.3.6 CONDITIONAL PRIMITIVES

Lisp provides several primitives for conditional execution. The simplest of these is IF. The IF primitive is described in a so-called *IF form*. In an IF form, the first *test form* argument determines whether the second argument, the *then form* (if the value of the test form is non-NIL) or the third *else form* argument (if the value of the test form is NIL) will be evaluated.

```
* (if (member 'b '(a b c)) 'member 'non-member)
MEMBER
```

```
* (if (null '(1 2 3)) 'empty-list 'non-empty-list)
NON-EMPTY-LIST
```

There are two special forms of the `IF` primitive which are as follows:

- In a `WHEN` primitive the else form is omitted. If the value of the test is `NIL` then nothing is done and the value of the `WHEN` form is `NIL`. Otherwise the return value is the value of the last argument.
- In an `UNLESS` primitive the then form is omitted. If the value of the test is non-`NIL` then nothing is done and the value of the `UNLESS` form is `NIL`. Otherwise the return value is the value of the last argument.

The use of the `WHEN` and `UNLESS` primitives is illustrated below:

```
* (when (member 'b '(a b)) (setf y '12) 'member)
MEMBER
* y
12

* (unless (member 'b '(a b)) (setf x 'x) 'non-member)
NIL
* x
ERROR (about that "x" has no value)
```

It is important to note that both `WHEN` and `UNLESS` can work with any number of arguments.

If we need more complicated conditions, we can use the `COND` primitive. The arguments of the `COND` primitive are so called *clauses.* The first element of a clause is a test followed by zero or more *consequences.* The `COND` form finds the first clause whose test form is evaluated to true (non-`NIL`) and executes all of its consequences and returns the value of the last consequence. The following two simple examples show the use of the `COND` primitive.

```
* (setf x 15)
15
* (cond ((not (numberp x)) 'not-number)
        ((> x 0) 'positive)
        ((< x 0) 'negative)
        (t 'zero))
POSITIVE

* (setf list '(a b c d))
(A B C D)
```

```
* (cond ((> (length list) 10) 'long-list)
        ((not (endp list)) 'short-list)
        (t 'empty-list))
SHORT-LIST
```

The LENGTH primitive counts the number of top-level elements in a list.

### 1.3.7 PROCEDURE DEFINITION

Some procedures supplied by Lisp itself are shown in the previous sections. However, users often need to define their own procedures, built from Lisp primitives and other user-defined procedures. The so-called user-defined procedures can be constructed with the help of the DEFUN primitive. The general form of the DEFUN primitive is the following:

```
(defun < procedure name >
      (< parameter1 > . . . < parametern >)
      < form1 >
      .
      .
      < formm >)
```

The first argument of the DEFUN primitive is a symbol indicating the name of the procedure, the second argument is a list of symbols, which contains the variable names that are used in the defined procedure. The body of the procedure contains the forms to be evaluated when the procedure is used. The return value of DEFUN is the name of the procedure, but its main purpose is to establish a procedure definition. The defined procedure can be used or called like any other procedure: with the expression consisting of the procedure name and its arguments.

---

**EXAMPLE 5.2** A procedure definition

In this simple example a procedure, which decides whether its argument is not a number or is a positive, negative or zero number, is defined.

```
* (defun number-check (x)
      (cond ((not (numberp x)) 'not-number)
            ((> x 0) 'positive)
            ((> x 0) 'negative)
            ((= x 0) 'zero)))
NUMBER-CHECK
* (number-check '(1 2 3))
```

```
NOT-NUMBER

* (number-check (* 1 -2 3))
NEGATIVE
```

---

# 1.4 SOME SIMPLE EXAMPLES IN LISP

The use of the Lisp language is illustrated with some simple examples in the following sections.

## 1.4.1 LOGICAL FUNCTIONS

Problem: Define the logical functions equivalence ($\equiv$) and implication ($\rightarrow$) with the help of the three basic logical predicates (AND, OR and NOT). The operation or truth tables of the logical functions are given in Table 5.1.

*Table 5.1.* Operation table of the "equivalence" and the "implication" operator

| $a \equiv b$<br>$a \downarrow \quad b \rightarrow$ | **nil** | **t** |
|---|---|---|
| **nil** | **t** | **nil** |
| **t** | **nil** | **t** |

| $a \rightarrow b$<br>$a \downarrow \quad b \rightarrow$ | **nil** | **t** |
|---|---|---|
| **nil** | **t** | **t** |
| **t** | **nil** | **t** |

Solution: The truth tables given in Table 5.1 show that the equivalence of two expressions is **t** if both of them are **nil** or both of them are **t**, and their implication is **t** when the condition part is **nil** or the consequent part is **t**. The equivalent Lisp description of the sentence above is as follows:

```
* (defun equivalence (a b)
      (or (and a b) (and (not a) (not b))))
EQUIVALENCE

* (defun implication (a b)
      (or (not a) b))
IMPLICATION
```

The use of the function above is illustrated by the following simple lines:

```
* (equivalence '(nil t))
NIL
```

```
* (equivalence '(nil nil))
T

* (implication '(nil t))
T

* (implication '(nil nil))
T
```

### 1.4.2 CALCULATING SUMS

Problem: Write a procedure that summarizes the elements of a list of numbers (a list containing numbers as its elements).

Solution-1: The first solution is rather simple. All we have to do is to add the symbol of the addition primitive ('+) to the beginning of the list and evaluate the list with the help of EVAL primitive.

```
* (defun sum (list)
      (eval (cons '+ list)))
SUM
```

We can use this procedure as follows:

```
* (sum '(2 3 4))
9

* (sum ())
0
```

Solution-2: The second solution is a recursive definition, where the solution is composed of the solution of the sub-problems. Namely, we could get the solution if we knew the sum of the rest of the list and added the value of the first element to this sum. But, we could get the sum of the rest of the list if we knew the sum of the rest of the rest of the list ... and so on. And if we have an empty list, its sum is zero. The above can be written in Lisp syntax as follows:

```
* (defun recursive-sum (list)
      (cond ((null list) 0)
            (t (+ (first list)
               (recursive-sum (rest list))))))
RECURSIVE-SUM
```

Its use is very simple, too.

```
* (recursive-sum '(2 4 6 8))
20
```

### 1.4.3 POLYNOMIAL VALUE

Problem: Define a procedure that calculates the value of a given polynomial in a given substitution value.

Solution: We shall prepare a recursive solution to the problem by algebraic transformation. The usual form of a polynomial can be transformed as follows:

$$P_n(x) = a_n + a_{n-1}x + a_{n-2}x^2 + ... + a_0x^n$$

$$P_0(x) = a_0$$
$$P_1(x) = a_0x + a_1$$
$$P_2(x) = (a_0x + a_1)x + a_2$$

.

.

$$P_i(x) = xP_{i-1}(x) + a_i \qquad i = 1, 2, ..., n$$

The transformation above is known as the Horner-arrangement, which shows that the value of the polynomial can be determined by recursive steps using our knowledge of the substitution value and the coefficient-list. In Lisp syntax we have:

```
* (defun Horner (x coefficient-list)
      (cond ((null (rest coefficient-list))
                   (first coefficient-list))
             (t (+ (first coefficient-list)
                   (* x (Horner x
                          (rest coefficient-list)))))))
HORNER
```

The following lines illustrate the use of the recursive procedure above.

```
* (Horner 2 '(5 4 3 2))
41
```

Of course, the coefficients equal to zero must appear in the coefficient-list, too.

```
* (Horner 4 '(0 8 0 -4 0 0 1))
3872
```

# 2. THE PROLOG PROGRAMMING LANGUAGE

The Prolog programing language has taken its name from Programming in Logic. It is rather a programming system in which first-order logic is used as a programming language. The first official version of the Prolog system was introduced in the early 1970s by Alain Colmeraurer at the University of Marseilles, France. Today Prolog is a very important tool in programming artificial intelligence applications and in the development of expert systems.

Prolog is a *declarative* programming language. This means that the user only needs to define the description of the problem and does not need to solve it. The solution is found by the Prolog interpreter in the form of an answer to a question with the help of logical reasoning. Thus, the fundamental differences between conventional programming languages and Prolog are as follows.

In conventional programming:

- The programmer defines an algorithm in the form of step by step instructions telling the computer how to solve the problem.
- The computer executes the instructions in the specified order.

In logical programming:

- The programmer defines the relationships between various entities with the help of logic.
- The system applies logical deductions to solve the problem.

## 2.1 THE ELEMENTS OF PROLOG PROGRAMS

While the basic functional notation in programming languages is the notation of mathematical functions, logical programming languages rely on the notion of relation. A Prolog program is a Prolog *database* composed of *relations* (or *predicates*). A predicate is defined by its name and by the number of its arguments. For example `likes/2` is a binary relation and `start/()` is a predicate with no argument. Each predicate is defined by one or more *clauses* in the program. This way a Prolog program is a description of a world with finite set of clauses, which can be either *facts* or *rules*. In this chapter the main elements of Prolog programs are described.

### 2.1.1 FACTS

The simplest form of Prolog predicates are the so called facts. Facts correspond to records in a relational database. They represent the statements or relations that are assumed to be true. Let us consider the facts below, for example:

| (Prolog form) | (explanation) |
|---|---|
| `toy(doll).` | "Doll is a toy." |
| `plays(ann,doll).` | "Ann plays with doll." |
| `father(john,ann).` | "John is the father to Ann." |
| `father(peter,john).` | "Peter is the father to John." |
| `lottery(10,[15,18,27,49,70]).` | "The lottery numbers of the 10. week are 15, 18, 27, 49 and 70." |
| `satisfied(X,X).` | "Everyone is satisfied with himself." |
| `person(name(ann), birthday(1990,may,12)).` | "The name of a person is Ann and her birthday is on 12 of May in 1990." |

Facts consists of:

- the predicate name such as `toy`, `plays`, `father`, `lottery`, `satisfied` and `person` (this must begin with a lower case letter),

- and zero or more arguments such as `doll`, `ann`, `john`, `peter`, `10`, `[15,18,27,49,70]`, `X`, `name(ann)` and `birthday(1990,may,12)`.

The syntactical end of facts and all Prolog clauses are denoted by a period.

The arguments can be any of the following Prolog *terms*:

- *atoms* such as `doll`, `ann`, `john`, `peter` and `may` represent indivisible specific part of the world and begin with lower case letter

- *numbers* such as `10`, `15`, . . ., `1990` and `12`

- *variables* such as `X` which represent an unspecified element and begin with an upper case letter or an underline character

- *structured objects* such as `name(ann)` and `birthday(1990,may,12)` which consist of a functor (e.g., `name`, `birthday`) and a fixed number of arguments, which can be any type of Prolog terms, too.

- *lists* such as `[15,18,27,49,70]` consist of a collection of terms, including structures and lists. Syntactically, a list is denoted by square brackets and the elements of the list are separated by commas.

The other symbols used in the facts above "(", ")", "." and "," are delimiters.

## 2.1.2 RULES

Rules represent things that are true depending on some conditions, for example:

| (Prolog form) | (explanation) |
|---|---|
| `likes(ann, X) :-` | "Ann likes every toy |
| `  toy(X), plays(ann, X).` | she plays with." |
| `child(X, Y) :-` | "X is the child to Y if |
| `  father(Y, X).` | Y is the father to X." |
| `sister(X, Y) :-` | "X and Y are sisters if |
| `  father(Z, X), father(Z, Y).` | they have the same father." |

A rule consists of a *head* and a *body*. For example the head of the first rule is `likes(ann,X)` and the body is `toy(X), plays(ann,X)`. The head of a rule is a predicate definition and the body is a set of conditions combined with a conjunction. The head and the body of a rule are separated by ":-" symbol which can be read as "if", and the parts of the body is separated by "," symbol which denotes logical "and".

Facts and rules are collectively called clauses, which essentially describe sentences. The order of clauses with different heads is optional in Prolog programs. Clauses with the same head are generally grouped into procedures and are tested in the order they appear in the program, from top to bottom.

## 2.1.3 QUESTIONS

The *question* or *goal* is used in Prolog programs to find out if something is true, for example:

| (Prolog form) | (explanation) |
|---|---|
| `?- toy(car).` | "Is car a toy?" |
| `?- likes(X, doll).` | "Who likes doll?" |
| `?- father(X, ann), father(Y, X).` | "Who is the father to Ann |
| | and the father of Ann's father?" |
| `?- person(name(ann), X).` | "When is Ann's birthday?" |
| `?- father(X, Y).` | "Who is the father to whom?" |

A goal can be a simple question consisting of only one predicate (e.g.: `?- toy(car).`) or more predicates can be combined to form a compound question (e.g.: `?- father(X,ann), father(Y,X).`). The answer given by Prolog is **yes** or **no** and the bindings of all variables in the question if they exist. So we might have:

```
?- toy(car).
no

?- likes(X,doll).
X = ann

?- father(X,ann), father(Y,X).
X = john
Y = peter

?- person(name(ann),X).
X = birthday(1990,may,12)

?- father(X,Y).
X = john
Y = ann;
X = peter
Y = john;
no
```

There are more than one solutions in the last example. In this case, the other possible bindings can be seen by typing ";" after Prolog prints out the first variable binding. The last `no` means there are no more solutions.

### 2.1.4 THE PROLOG PROGRAM

In Prolog programs a special class of the first order logic, the so-called *Horn clause* is used. A Horn clause or Horn sentence has the following form:

$$A \leftarrow B_1 \wedge B_2 \wedge \ldots \wedge B_n$$

with Prolog notation:

$$\texttt{A :- } B_1, B_2, \ldots, B_n.$$

where A and $B_i$'s are predicates.

There are three possible types of Horn clauses conventionally named as follows:

- a clause of the form "`A.`" is called a *fact* (facts have head but no body)
- a clause of the form "$A \leftarrow B_1 \wedge B_2 \wedge \ldots \wedge B_n$" or with Prolog notation "$\texttt{A :- } B_1, B_2, \ldots, B_n.$" is called a *rule* (rules have both head and body)

- a clause of the form "$B_1 \wedge B_2 \wedge \ldots \wedge B_n$." or with Prolog notation "`B1, B2, . . ., Bn.`" is called a *goal* (goals have body, but no head)

A Prolog program consists of facts, rules and goals together.

---

**EXAMPLE 5.3** A simple Prolog program

```
likes(ann,X) :- toy(X), plays(ann,X).
toy(car).
toy(doll).
plays(ann,doll).

?- likes(ann,What).
```

---

### 2.1.5 THE DECLARATIVE AND PROCEDURAL VIEWS OF A PROLOG PROGRAM

The two interpretations of the Prolog language form the speciality of Prolog and logical programming.

The *declarative* reading of the clause

$$A :- B_1, \ldots, B_n.$$

is: "A is true if $B_1$ is true and . . . and $B_n$ is true". So Prolog statements are translated as logical forms and the answer to a question is a set of substitutions, which can be used for the deduction of the question from the statements. The declarative meaning is able to make the programs more readable, because it is only a small separate part of the program that has to be interpreted at the same time.

The *procedural* reading of the clause above is: "To solve problem A, first solve problem $B_1$, then solve problem . . . and then solve problem $B_n$". So the procedural interpretation gives the algorithm of execution, in other words it shows how a given problem can be solved.

### 2.1.6 MORE ABOUT LISTS

As it was mentioned in Section 2.1.1 of this Chapter, a list is a collection of zero or more terms such as atoms, numbers, variables, structured objects and other lists. There is a special list, the *empty list*, which is denoted by a pair of square brackets: `[]`.

A list is a recursive data structure. As in the Lisp language, lists in Prolog also consist of two parts: the *head*, which is the first element, and the *tail* which must be a list, too, containing the remainder part of the list. For example,

the head of `[1,2,3]` is `1` and the tail is `[2,3]`;
the head of `[a(1,2),a(3,4)]` is `a(1,2)` and the tail is `[a(3,4)]`;
the head of `[[a,b]]` is `[a,b]` and the tail is `[]`.

There is a special notation for list structures: instead of separating elements with commas, the head and the tail can be separated with a vertical bar "|". For example,

`[1,2,3]` is equivalent to `[1|[2,3]]`,
which is equivalent to `[1|[2|[3]]]`,
which is equivalent to `[1|[2|[3|[]]]]`

In Prolog, the head and the tail of a list can be selected by pattern matching the actual list with the notation [X|Y], where the head of the list is bounded to X and the tail of the list is bounded to Y. For example,

in case of `[1,2,3]` `X=1` and `Y=[2,3]`;
in case of `[a(1,2),a(3,4)]` `X=a(1,2)` and `Y=[a(3,4)]`;
in case of `[[a,b]]` `X=[a,b]` and `Y=[]`.

The pattern matching mechanism of Prolog and this special notation for list structures enables the dissection and the construction of lists.

## 2.2 THE EXECUTION OF PROLOG PROGRAMS

The execution of a Prolog program aims to prove the goal and find the value for the variables, using a *built-in theorem proving algorithm*. In the following sub-sections, the operation of a Prolog program and the main characteristics of this algorithm are shown in detail.

### 2.2.1 HOW QUESTIONS WORK

Let us now examine how Prolog answers a question with the help of the simple Prolog program in Example 5.3. We have the goal:

```
?- likes(ann,What).
```

Prolog tries to prove the question by looking for facts which match this goal, or rules whose heads match this goal and whose body can be proved. Evaluation steps:

1. The clause

```
likes(ann,X) :- toy(X), plays(ann,X).
```

   is found and matched with the goal. The unifier is the substitution `What|X`, and the body of the rule becomes a new goal. So we have two new subgoals: `toy(What)` and `plays(ann,What)`.

2. Now, to evaluate the first subgoal, the system finds the fact

```
toy(car).
```

   and unifies the variable `What` and the constant `car`.

3. After matching, the second subgoal becomes `plays(ann,car)`. It is not unifiable with any fact and with the head of any rule in the program. In this case the system must go back to a preceding subgoal and needs to find another possible alternative.

4. There is another fact in the program matching the subgoal `toy(What)`:

```
toy(doll).
```

   The unification is `What|doll` and the second subgoal becomes to `plays(ann,doll)`.

5. The second subgoal is unifiable with the fact `plays(ann,doll)`. There are no more subgoals, so goal evaluation has succeeded and the system returns with the answer: `What = doll`.

As you have seen in this simple example, the two main mechanisms of the theorem proving algorithm are *pattern matching* or *unification* and *backtrack*.

The search tree (an AND-OR tree, mentioned in Section 3.1 of Chapter 3) traversed during determination of the response of Prolog in Example 5.3 is illustrated in Fig. 5.4. The arcs of the tree denote the response of the subgoals. The root contains the goal and the subgoals deriving from the initial goal can be found in the other nodes. The number of hyperarcs originating from a node is equal to the number of answers of the first subgoal. The leaf nodes include the subgoals matching with a fact of the Prolog program and the cases when the subgoals cannot be proved.

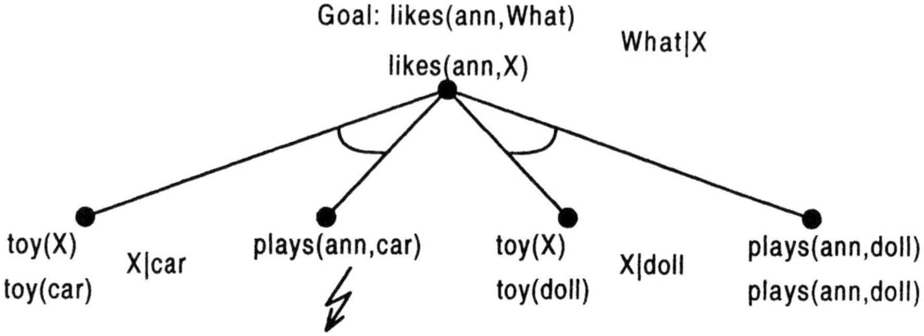


*Figure 5.4.* The search tree of the question

### 2.2.2 UNIFICATION

Parameters are passed on using *bidirectional pattern matching* or *unification* in Prolog. During unification the subgoal and the head of the clause must have the same uniform structure with substitutions of variables.

The conditions of unification are the following:

- the predicates have the same name
- the predicates have the same number of arguments
- the arguments are unifiable as follows
  - a variable and any term is always unifiable
  - two primitive terms (atom or number) only unify if they are identical
  - two structures unify if they have the same functor and the arguments are unifiable one after the other

Let us examine some examples to illustrate the condition of unification:

Case1 : $\mathtt{p(1,b,d)}$
$\mathtt{q(2,B,B,D)}$

The predicates are not unifiable as the names of the predicates are not equivalent.

Case2 : $\mathtt{p(1,b,d)}$
$\mathtt{p(2,B,B,D)}$

The matching is not successful as the argument numbers are different.

Case3 : `p(1, b, d)`
`p(2, B, B)`

The names and the argument numbers of the predicates are the same, but the first arguments are not unifiable, because both of them are numbers with different values.

Case4 : `p(1, b, d)`
`p(1, B, B)`

The first and the second arguments are unifiable with the binding B|b, but the third arguments (d and B|b) cannot be matched.

Case5 : `p(1, b, d)`
`p(1, B, D)`

The unification is successful with the matching list: B|b, D|d.

The role of unification is dual:

- the clause applicable to the subgoal is selected by pattern matching

- and parameter passing is also performed by the proper variable-substitution in the unification step.

### 2.2.3 BACKTRACKING

As you can see in section 2.2.1 of this Chapter in step 3, when a subgoal fails in Prolog, the system backtracks to a previous subgoal to find an alternative possibility for the solution.

Backtracking has the following preconditions:

- the solution of a subgoal is not successful

- there are more solutions of a previously satisfied subgoal

- there is an untested possibility

A simple illustration of Prolog's backtrack mechanism is shown in Fig. 5.5. Considering a compound goal $G_1$,$G_2$,$G_3$. Assume that the first subgoal $G_1$ has been successfully executed and the second subgoal $G_2$ is being proved. Suppose that the subgoal $G_2$ unifies with the head of the clause C:-$P_1$,$P_2$,$P_3$ and the subgoals $P_1$ and $P_2$ are satisfied. When $P_3$ fails, the system goes back to subgoal $P_2$ and tries the other untested possibility. If $P_2$ also fails, than it can go back to $P_1$ and when this subgoal fails, too, it goes back to the next clause which unifies with $G_2$, and so on.

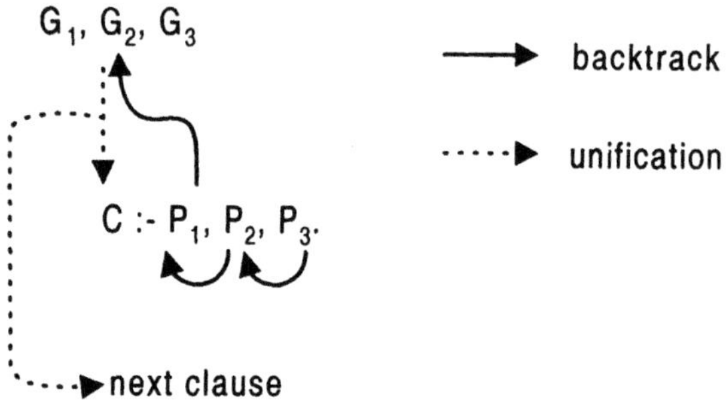


*Figure 5.5.* Backtracking in Prolog

### 2.2.4 TRACING PROLOG EXECUTION

The best way to understand Prolog execution is the use of a tracing facility based on the basic control flow model in Fig. 5.6. Prolog tells us when

- it *calls* a clause,
- it *exits* a clause successfully,
- a clause *fails*,
- it *retries* a clause because of backtracking.

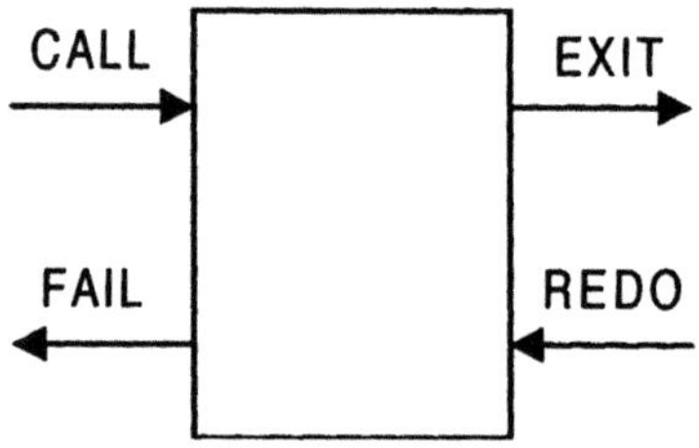


*Figure 5.6.* The control flow model of Prolog

The state of the Prolog inference engine and its actions in the four states above are the following:

- call: Prolog begins searching for clauses that unify with the subgoal.
- exit: The subgoal is satisfied and the appropriate variables are bound.
- fail: This state indicates that no more clauses match the subgoal.

- retry or redo: This indicates backtrack, when Prolog unbinds the variables and retries the subgoal.

---

**EXAMPLE 5.4** Example 5.3 (continued)

Let us see the execution steps of the simple Prolog example 5.3:

```
?- likes(ann,What).
CALL: likes(ann,What)

    CALL: toy(What)
    EXIT: toy(car)

    CALL: plays(ann,car)
    FAIL: plays(ann,car)

    REDO: toy(What)
    EXIT: toy(doll)

    CALL: plays(ann,doll)
    EXIT: plays(ann,doll)

EXIT: likes(ann,doll)
What=doll
```

---

### 2.2.5 THE SEARCH STRATEGY

The simple examples in the earlier sections of this Chapter showed how to answer a Prolog question. Let us summarize what we have learned in the following points:

1. Prolog does backward chaining with depth-first search.
2. The order of subgoals determines the sequence in which subgoals are satisfied (left to right).
3. The clauses are tested in the order they appear in the program (from top to bottom).
4. When a subgoal matches the head of a rule, the body of that rule must be satisfied as a new set of subgoals.
5. A goal has been proved when all of its subgoals are satisfied.

### 2.2.6 RECURSION

In almost any Prolog program you can find *recursive* clauses - clauses that call themselves. In a recursive clause the predicate symbol of the head occurs as a predicate symbol in the body, too.

In any language, a recursive definition consists of at least two parts:

- the trivial case that is known to be true,
- the reduction of the general case to the trivial case.

The same principle holds for recursion in Prolog as it is illustrated by the simple example below.

---

**EXAMPLE 5.5** A simple recursive example

Suppose we want to define a Prolog definition to determine whether there is a path from a node to another node in a directed graph. The problem can be defined as follows:

- there is a path from X to Y if there is an arc from X to Y (the trivial case),
- there is a path from X to Y if there is an arc from X to Z and there is a path from Z to Y (the reduction).

This can be written in Prolog as follows:

```
path(X,Y) :- arc(X,Y).

path(X,Y) :- arc(X,Z), path(Z,Y).
```

The program is to be completed with a list of facts giving the arcs of the graph.

---

## 2.3 BUILT-IN PREDICATES

Prolog includes several built-in predicates for arithmetic manipulations, input/output, and various other system and knowledge base functions. Some of there predicates are summarized in the following sections.

### 2.3.1 INPUT-OUTPUT PREDICATES

Different Prolog expressions can be written to and read from the console or file with the help of built-in input-output predicates. For example, the predicate `write/1` writes the current value of its argument to the current output device, the predicate `nl/()` generates a new line and `read/1` reads a term from the current input device and unifies it with its argument. Normally, the current input device is the keyboard, and the screen is used for output.

```
?- write('Hello!').
Hello!

?- write([1,2,3]).
[1,2,3]

?- nl.


?- read(X).
ann.
X=ann

?- read(Hour:Min).
8:10.
Hour=8
Min=10
```

### 2.3.2 DYNAMIC DATABASE HANDLING PREDICATES

Prolog allows us to manipulate, i.e. to add and remove clauses in the program. The modifiable predicates are called *dynamic predicates* and have to be declared as dynamic. In order to add new clauses to a database, the built-in predicates `asserta/1` and `assertz/1` (or shortly `assert/1`) are used, and they cause the new clause to be inserted before the first and after the last clause of the predicates with the same head. In order to remove a clause from a database, the predicate `retract/1` is used.

```
?- assert(plays(ann,doll)).
yes

?- asserta(plays(john,car)).
yes

?- plays(X,Y).
X=john
```

```
Y=car;
X=ann
Y=doll;
no

?- retract(plays(john,car)).
yes

?- plays(john,X).
no
```

As you have seen in the examples in the previous sections, there are no global variables in Prolog, the form of the Prolog database is used for those, too. Information can be stored in facts and can be manipulated with `asserta`, `assert` and `retract`.

### 2.3.3 ARITHMETIC PREDICATES

The arithmetic predicates (e. g. <, <=, `is`) and the arithmetic functions (e. g. +, -, *, /) are used for the evaluation and comparison of arithmetic expressions.

```
?- 3 < 2*5.
yes

?- 4-1 > 9/3.
no

?- X is 3+4.
X=7

?- 10 is 5*2.
yes
```

### 2.3.4 EXPRESSION-HANDLING PREDICATES

Expression-handling predicates are used for taking apart and connecting Prolog expressions. For example, the predicate `append/3` concatenates lists and the predicate `concat/3` combines its first and second arguments to form the third argument.

```
?- append([1,2],[3,4],X).
X=[1,2,3,4]

?- append(X,Y,[a,b]).
X=[]
Y=[a,b];
X=[a]
Y=[b];
```

```
X=[a,b]
Y=[];
no

?- concat(moon,flower,X).
X=moonflower

?- concat(life,X,lifetime).
X=time
```

### 2.3.5 CONTROL PREDICATES

Section 2.2.5 in this Chapter describes the search strategy of Prolog with an explanation of how the order of goals and clauses affect the execution of a program. In this section we will show two main techniques which are used to control the search mechanism in Prolog: the `fail/()` predicate, which is used to force backtracking, and the `!` (cut) predicate, which is used to prevent backtracking.

Recall that when the evaluation of a subgoal fails in Prolog, the system backtracks to a previous subgoal to find an alternative solution. In certain situations it is necessary to force backtracking in order to seek out more or even all of the possible solutions. Prolog has a built-in predicate, `fail`, which represents a subgoal that is never satisfied (it always fails), so Prolog is forced to backtrack.

One of the most important control predicates is `cut`, which is represented by an exclamation mark "`!`". The effect of the `cut` operation is very simple: it always succeeds, but it is impossible to backtrack across the `cut`. The name of the predicate indicates the cutting of the search tree. The backtrack nodes which would be executed after calling `cut` are simply omitted by it. The effect of cut is shown in Fig. 5.7.

## 2.4 SOME SIMPLE EXAMPLES IN PROLOG

The use of the Prolog programming language is illustrated by some simple examples in the following sections.

### 2.4.1 LOGICAL FUNCTIONS

Problem: Define the equivalence and implication logical functions.

Solution: The operation or truth table given in Table 5.1 in section 1.4.1 of this Chapter shows that the equivalence of two expressions is true when their logical values are the same, and the implication is true when the condition part is false or the consequent part is true.

The reasoning above is the following coded in a Prolog form.

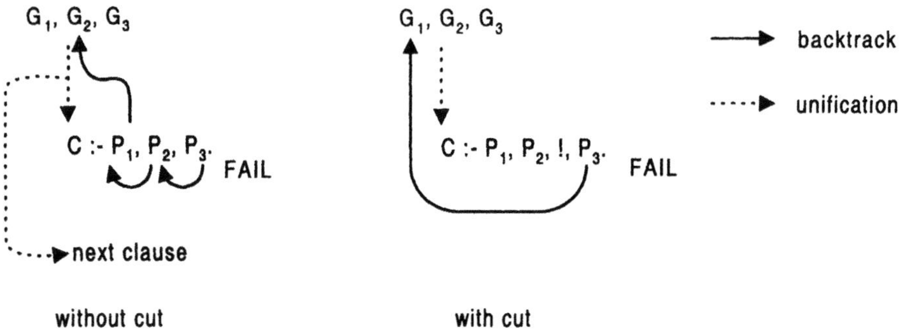


*Figure 5.7.* The effect of cut

```
equivalence(X,X).

?- equivalence(false,true).
no

?- equivalence(false,false).
yes

implication(false,_) :- !.
implication(_,true).

?- implication(false,true).
yes

?- implication(false,false).
yes
```

### 2.4.2 CALCULATION OF SUMS

Problem: Define a Prolog program that calculates the sum of the integers which lie in between two given integer numbers.

Solution: A recursive program is used for the solution:

```
summarize(Less,Bigger,Sum) :-
      Less<Bigger,!,
      recursive_sum(Less,Bigger,Less,Sum).

summarize(Bigger,Less,Sum) :-
      recursive_sum(Less,Bigger,Less,Sum).

recursive_sum(Less,Less,Sum,Sum) :- !.

recursive_sum(Less,Bigger,Aux,Sum) :-
      Less<Bigger,
```

```
        New_Less is Less+1,
        New_Aux is Aux+New_Less,
        recursive_sum(New_Less,Bigger,New_Aux,Sum).
```

The clause **summarize** determines which of the arguments is a smaller number and starts the **recursive_sum**, also consisting of an auxiliary parameter equal to the smaller number. In the first, trivial case of the **recursive_sum** clause, when the smaller and the bigger numbers are the same, the sum is equal to the auxiliary parameter. Adding the smaller number to one and adding the auxiliary parameter to the new smaller number, the procedure **recursive_sum** is called again with the new arguments in the second case of the clause.

This program can be used as follows:

```
?- summarize(52,128,X).
X=6930

?- summarize(128,52,6940).
no
```

### 2.4.3 PATH FINDING IN A GRAPH

Problem: Let us consider the path finding program mentioned in Example 5.5:

```
path(X,Y) :- arc(X,Y).

path(X,Y) :- arc(X,Z), path(Z,Y).
```

and consider a directed graph shown in Fig. 5.8.

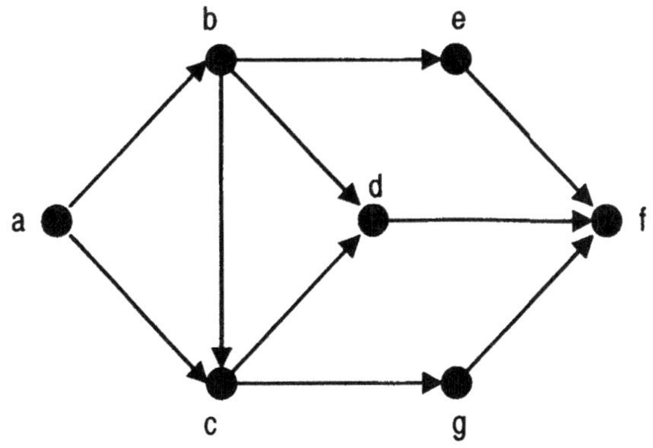


*Figure 5.8.* A directed graph

1. Define the Prolog description of the directed graph in Fig. 5.8 and examine the behaviour of the program.

2. Change the direction of the arc b→d in the graph to d→b and examine the behaviour of the program again. Modify the Prolog definitions in order to be able to handle the altered graph.

Solution:

1. Arcs can be defined by the description of Prolog facts as follows:

```
arc(a,b).
arc(a,c).
arc(b,c).
arc(b,d).
arc(b,e).
arc(c,d).
arc(c,g).
arc(d,f).
arc(e,f).
arc(g,f).
```

   We can test the behaviour of the program by giving some questions as follows:

```
?- path(a,f).
yes

?- path(f,a).
no

?- path(c,X).
X=d;
X=g;
X=f;
X=f;
no
```

   Node `f` appears twice in the solution as it is reachable by two different paths from node `c`.

2. As the modified graph consists of a directed cycle, the Prolog program can get into an endless loop. So the travelled nodes must be noticed if we want to avoid endless running. We can revise the clauses of `path` as follows:

```
path(X,Y,Nodes) :-
        arc(X,Y),
        not(member(Y,Nodes)).

path(X,Y,Nodes) :-
        arc(X,Z),
        not(member(Z,Nodes)),
        path(Z,Y,[Z|Nodes]).
```

   The clause `member` is used for examining whether its first argument is an element of the list in its second argument.

   Now we might have the question:

```
?- path(X,d,[]).
X=c;
X=a;
X=b;
X=a;
X=d;
no
```

with the correct answer.

## 3. EXPERT SYSTEM SHELLS

In section 2. of Chapter 1 we have already introduced the notion of knowledge-based systems. They are computer systems that contain stored knowledge and solve problems like humans would. *Rule-based expert systems* are knowledge-based systems which are applied in a narrow specific field and possess a rule-based knowledge base. They solve difficult problems which would require a specialized human expert. In particular, they can make intelligent decisions and can offer intelligent advice and explanations.

Roughly speaking, expert system shells are "empty" expert systems in the sense that they contain all the active elements of an expert system, but the special, domain specific knowledge is missing from the knowledge base. The basic components of an expert system shell are shown in Fig. 5.9 in the dotted box.

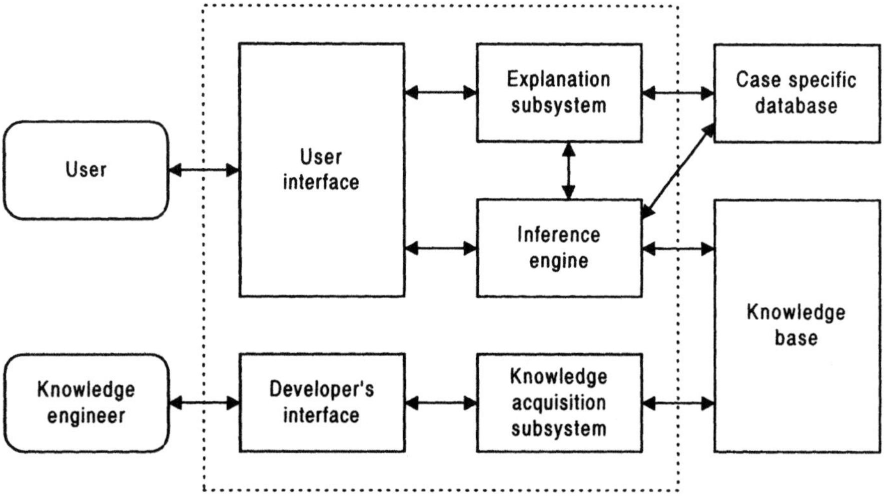


*Figure 5.9.* The components of an expert system

## 3.1 COMPONENTS OF AN EXPERT SYSTEM SHELL

Fig. 5.9 as a whole depicts the structure of an expert system composed of several basic components as follows:

- *knowledge base*
  As it has been discussed before in detail in Chapter 2, the knowledge base stores the factual and heuristic knowledge in any expert system and is one of the standard components.

- *case specific database*
  The task specification(s) to be solved by the expert system are in this database.

- *inference engine*
  The inference engine is also a standard element in an expert system. It manipulates the symbolic information and knowledge in the knowledge base to perform reasoning when solving a problem. Chapter 3 deals with reasoning in details.

- *user interface*
  The user interface is a standard component in almost any software system. It allows the user to interact with the system in an easy "user-friendly" way.

- *explanation subsystem*
  The explanation subsystem is a service utility which explains the system's actions upon the request of the user.

- *knowledge acquisition subsystem*
  The knowledge acquisition subsystem is also a service utility, the expert system counterpart of a database management utility. It is used for updating, checking, verifying and validating the knowledge base (see Chapter 4 for details).

- *developers' interface*
  A software developers' interface can be found in almost any software system as a standard component. In the case of expert systems, it allows the knowledge engineer to interact with the knowledge acquisition subsystem.

Some of the components, such as the knowledge base, the knowledge acquisition subsystem and the inference engine, are subjects of earlier chapters.

The dotted box in Fig. 5.9 encapsulates the components of an *expert system shell*, which is an environment for creating expert systems with different domain specific knowledge. The components of an expert system shell include the non-specific parts of an experts system, so it contains the inference engine, the explanation subsystem, the knowledge acquisition subsystem and the interfaces. An expert system shell can be imagined as an "empty" expert system with a powerful developer's subsystem.

## 3.2 BASIC FUNCTIONS AND SERVICES IN AN EXPERT SYSTEM SHELL

The basic, commonly used functions and services offered by an expert system shell are briefly described here. They are arranged according to the shell-specific components above.

1. *Explanation functions and services*
   The explanation subsystem of an expert system shell offers the following functions to help the user to follow and understand the reasoning process and the reasoning results provided by the system:

- explanative reasoning
  which provides an answer to the questions:
  *"Why?"* and *"How?"*

  Besides the result of reasoning, the explanation function gives information to the user about the way the result has been found. Consequently, these functions are closely related to the tracing functions provided by the knowledge acquisition subsystem.

  The answer to the question *"Why?"* consists of the knowledge elements used for deriving the reasoning result. The full tracing information about the actual reasoning steps is in the answer to the question *"How?"*.

- hypothetical reasoning
  which gives an answer to questions of the type
  *"What would happen if* **eq-expr** *were true?"*

  where **eq-expr** is a value assignment statement, like "$X = 2$" with $X$ being a variable.

  Hypothetical reasoning involves a conditional assignment present in **eq-expr** and the derivation of its consequences by reasoning and/or simulation. It is important to note that the assignment and its consequences can be withdrawn if the user is not satisfied with it.

  The presence of the hypothetical reasoning function assumes the existence of an inference engine with a conditional reasoning facility and a simulator in the case of real-time expert systems.

- counterfactual reasoning
  This service function searches for counter-examples of a logical statement within the actual content of the knowledge base.

2. *Knowledge acquisition tools and services*
   The knowledge acquisition functions are offered by the knowledge acquisition subsystem of an expert system shell through the developers' interface. The primary user of these functions is the knowledge engineer, who is a person with high qualification, trained in expert systems. The following functions are usually offered.

   - checking the syntax of the knowledge element(s)

   - checking the consistency of the knowledge base
     This is a complex function for the verification and validation of the entire content, which usually includes the test(s) for contradiction freeness and completeness (described in Chapter 4).

- knowledge extraction
  to collect information from the knowledge base that satisfies the criteria defined by the knowledge engineer (the extraction filter).
  The result of the extraction may be used for verification and/or validation purposes or may be exported to another knowledge or database.
- automatic logging or book-keeping of the changes to the knowledge base,
  which is useful for tracing and maintenance purposes. This function can also be useful for repairing the knowledge base when combined with hypothetical reasoning and consequence withdrawal in the case of any consistency problems.
- tracing facilities
  These group of service functions includes general functions, such as
  - specification and handling of breakpoints for the reasoning process
  - automatic monitoring and reporting of the change in the values of the knowledge elements during reasoning (also combined by breakpoint generation)

  implemented in an expert system environment.

3. *Interface functions and services*
   According to the interfaces an expert system may have, the following function groups are available:
   - user and developer interfaces
   - operating system interface
   - real-time data exchange interface (if applicable)
     This interface function group plays a central role in real-time expert systems, which are the subject of Chapter 6.

It is important to note that a specific implementation of the general functions and services of an expert system shell can be observed on the example of the G2 real-time expert system shell in Chapter 10.

# Chapter 6

# REAL-TIME EXPERT SYSTEMS

This chapter summarizes the software architecture and properties of real-time expert systems. The material in the chapter is used extensively later in Chapter 10 where a concrete example of a real-time expert system shell, the G2 real-time expert system shell is described.

As we have seen in Chapters 2 and 3, expert systems exhibit unusual specific properties both in their data and procedural elements as compared to conventional software systems. In addition to this, real-time systems are special software systems which should serve special purposes and as a result have special properties. Therefore, the real-time and intelligent components are usually implemented in separate environments as relatively autonomous subsystems. Special attention is then paid to the cooperation and coordination of the real-time and intelligent elements in an intelligent control system [46].

In accordance with the key issues in a software architecture of a real-time expert system, the chapter is broken down into the following sections.

- The architecture of real-time expert systems [47]
- Synchronization and communication between real-time and intelligent subsystems
- Data exchange between real-time and intelligent subsystems
- Software engineering of real-time expert systems

# 1. THE ARCHITECTURE OF REAL-TIME EXPERT SYSTEMS

The software components of a real-time expert system are those of an expert system with the necessary real-time software elements added [47], [48]. The two types of components are usually separated by an interface through which data exchange and synchronization is carried out as shown in Fig. 6.1 below.

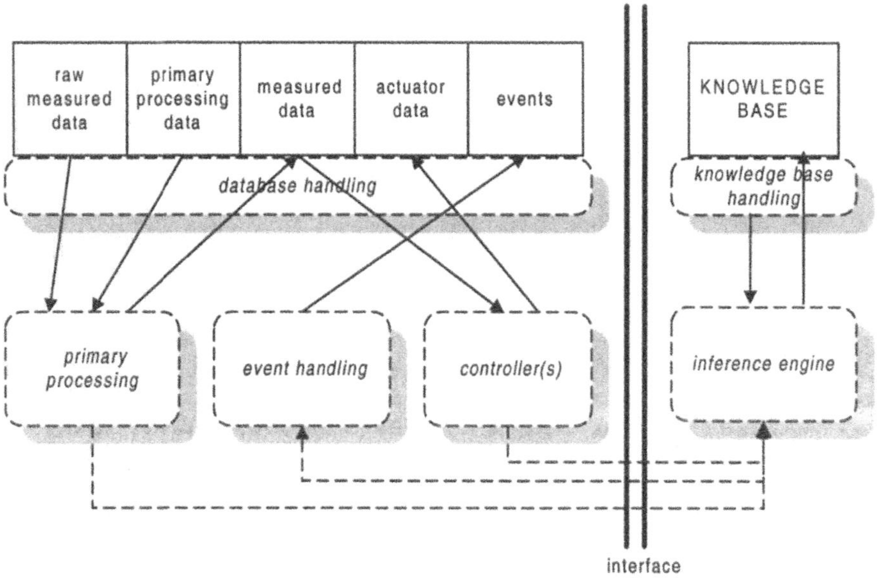


*Figure 6.1.* The architecture of real-time expert systems

The passive elements, that is the data files in a database and the knowledge base are denoted by squares. The active elements including tasks or processes and the data- or knowledge base manager are depicted in rounded dashed squares. Data exchange between the active and passive elements is shown by arrows, and the synchronization and communication between active elements by dashed arrows.

The components of the expert system part located on the right side of the figure form the so called *intelligent subsystem*, the other components on the left side responsible for the real-time control behaviour are called *real-time subsystem*. The interface denoted by bold double line in the middle clearly separates and connects the two subsystems.

In the remaining part of this section the most important real-time and intelligent components are briefly introduced and described.

## 1.1 THE REAL-TIME SUBSYSTEM

The real-time subsystem is usually a real-time control system with the same or similar software components found in a computer controlled system. The software architecture of a computer controlled system is described and discussed in detail in section 5. in Appendix A.

There are some key properties that a real-time system, including intelligent control systems, should possess.

1. time-dependent reactions
   Absolute and relative time is an important synchronization mechanism in real-time systems. There are periodic and time-dependent tasks to be carried out in relation to control and monitoring functions which are driven by a special hardware or software clock that belongs to the kernel of the operating system.

2. finite prescribed response time
   A characteristic requirement for a real-time system is that it should perform any required response in a prescribed finite time. Therefore none of its functions should include waiting for a possibly infinitely long or unpredictable time period.

3. time-out
   To avoid waiting too long for something, there is a special mechanism, called *time-out*. It cancels the action if a prescribed time interval is over, and performs some default action in response together with a warning message and resets all pending related actions if necessary.

4. no loss of raw data
   The load of a real-time system can be measured in terms of the number of changes in the signals in unit time that affect the system behaviour and require response or action from it. The load of a given real-time system typically varies in several orders of magnitude in time: there are low-load periods when almost nothing happens, and then in case of emergency situations the load may increase to a hundred or thousand times of the average. This highly varying nature of the load implies that almost all real-time systems are designed for some kind of overall load, or a bit higher. Special mechanisms take care of the behaviour of the overloaded system in high load periods.

   One of the requirements of real-time systems is that raw data should not be lost even during highly overloaded periods. The system may not be able to process all of data but it should store all the received signal changes for possible processing later.

5. priority handling
   One way a real-time system (and everyone) cops with overload is that the most important tasks are done first at the price of neglecting the others. This policy requires priorities to be set for every possible task or process in a real-time system and a mechanism to handle priorities and control system behaviour according to these. Usually there is a combination of fixed and time-varying priorities in each real-time system.

6. "nice degradation"
   There can be cases when the overload persists, then the system should restrict its activity to the most important actions with the highest priority. Such a case is clearly a degradation of system performance from the viewpoint of its users. It is usually required that such an unavoidable degradation should be "nice" in the sense that it should allow the user to perform the most basic tasks to get information about the system itself and about the signals. Careful software design is essential for the implementation of a nice degradation. One way to achieve this goal for example is the application of advanced conditional and time varying priorities.

From the viewpoint of real-time expert systems is important to notice that not all of the essential elements of a real-time system will be connected to the elements of the intelligent subsystem. Figure 6.1 only shows the key elements to be interfaced with the intelligent subsystem:

- *primary processing*
  which may require intelligent diagnosis and/or prediction when anomalies in raw measured data are found,

- *event handling*
  to record every event, i.e. findings, results, abnormalities etc. in the system including the notification of the real-time subsystem about the results of the intelligent subsystem,

- *controllers* in wide sense
  which perform control, regulation, diagnosis and identification of the plant to be controlled. They plan and execute actions via setting actuator values and/or informing the operator about fault detection and isolation results, diagnostic findings, predictions etc. These high level tasks may require intelligent steps to be performed using the services of the intelligent subsystem.

## 1.2 THE INTELLIGENT SUBSYSTEM

The components of an expert systems have already been briefly introduced in section 2. of Chapter 1. The non-service elements of an expert system relevant in the context of real-time expert systems are shown in the right hand side of Fig. 6.1. Since expert systems are special knowledge based systems, they have to contain the following elements:

- knowledge base

- inference engine

- knowledge base manager.

Notice that the connections between these non-service elements and the real-time subsystem are also shown in Fig. 6.1. Observe that the intelligent subsystem only reads but does not write in the real-time database files. The result of reasoning is communicated to the real-time processes via task-task communication messages.

In order to understand the challenges of interfacing real-time and intelligent software components [49], we briefly recall some of the most important properties that characterize the knowledge and reasoning of an expert system.

a. *The knowledge (data) elements in a knowledge base are strongly related.*
This implies that part of the knowledge cannot be suitably organized into files that are locked separately when used in write mode, as is usually done in real-time databases. Instead, the whole knowledge base should be locked for the inference engine when it performs a reasoning task.

b. *Reasoning is computationally hard.*
This means that we cannot give a definite upper limit for the time needed to perform a reasoning task, this may strongly vary with the actual reasoning task and may well exceed the overall time-out value of the real-time subsystem. Therefore "loose" communication is to be implemented between the real-time and the intelligent subsystems, where the real-time part should not be waiting without for the result of the reasoning without its given time-out period.

# 2. SYNCHRONIZATION AND COMMUNICATION BETWEEN REAL-TIME AND INTELLIGENT SUBSYSTEMS

As a consequence of the separation and relative autonomy of the intelligent and real-time software elements, there is a need to organize their cooperation. The synchronization and communication between the real-time and intelligent subsystems are implemented as functions of the interface element (see in Fig. 6.1). The figure shows that both data (knowledge) exchange and synchronization are taking place between the two subsystems.

This section is devoted to the synchronization and communication between the active elements, that is processes or tasks of the two subsystems. More precisely we shall investigate the possible ways real-time processes, primary processing and controllers communicate their request for reasoning to the inference engine and the engine's response.

## 2.1 SYNCHRONIZATION AND COMMUNICATION PRIMITIVES

There are in principle four primitives, that is elementary operations, used in the synchronization and communication between processes:

ss: *send signal* and does not wait for acknowledgement

sa: *handshake*, i.e. sending a signal and waiting for an acknowledgement

ms: *send message* and does not wait for acknowledgement

ma: *message exchange*, i.e. sending and receiving messages

Primitives "ss" and "sa" are used for synchronization where only a single bit of information "something has happened" is exchanged, whereas primitives "ms" and "ma" are used for communicating messages. Almost every operating system, especially multitasking and/or real-time ones, offers support for process-process synchronization and communication in the form of e.g. semaphores and mailboxes.

When organizing the synchronization and communication of processes in a real-time expert system, we have to remember that a loose connection is needed between the elements of the intelligent and the real-time subsystem in order to meet the real-time requirement of finite prescribed response time. Therefore only the primitives that do not wait for acknowledgement - that is "*send signal*" and "*send message*" - are to be used. The fully synchronized types, that is the "***handshake***" and "***message exchange***", can be implemented in a loose way by using two one-way primitives of the appropriate type.

For communication purposes between the real-time processes and the inference engine, most often only the "*send message*" primitive is needed and is used in the interface of real-time expert systems. The messages are collected into separate *message queues* for each communicating pair and communication direction. This way we may have a queue of messages, i.e. reasoning requests, from primary processing to the inference engine, another one for the reasoning result messages from the inference engine to a particular controller etc. (compare with the dashed arrows in Fig. 6.1). The messages in a queue have a time stamp indicating the time of the request for reasoning results and are usually processed in a FIFO (first in first out) order.

The real-time processes are implemented in such a way that they are able to wait for a message but it is not a standard feature of an inference engine. *Therefore inference engines in a real-time expert system should be implemented in a way to be able to handle message queues appropriately.*

## 2.2 PRIORITY HANDLING AND TIME-OUT

Both priority handling and the presence of a time-out mechanism are basic requirements for real-time systems, therefore they are naturally required in real-time expert systems, too.

In order to meet the two requirements above, both the interface and the inference engine should meet additional criteria.

The simplest way to implement *priority handling in real-time expert systems* is to associate priorities with each of the reasoning request messages. The priority of the request is passed on to the priority of the reasoning result message.

From the viewpoint of the inference engine priority handling means the processing of the reasoning request message with the highest priority in the waiting queues. In the case some messages have the same priority, the inference engine follows FIFO order when processing.

In order to avoid having to scan all the incoming message queues for high priority messages, we can use one incoming queue for each priority class. In this case, real-time processes put their reasoning request message into the corresponding queues together with a time stamp and the identifier of the sender.

In every case, however, *suitable additional mechanisms, which take care of priority handling of incoming reasoning request messages and of outgoing reasoning result messages ought to be present in the inference engine of a real-time expert system.*

The *time-out mechanism* is also a natural part of the real-time subsystem. The time-out requirement, however, does not immediately imply that a corresponding time-out mechanism is also present in the intelligent subsystem, because the two subsystems are loosely coupled and intelligent processing is driven by messages and message queues. Any real-time task requesting reasoning to be performed will wait for the result only for the specified time-out interval, thereafter it will reset itself. If the result arrives late it will wait in the incoming queue of the real-time process till the result of the next request arrives in time, then it will be discarded and only the valid result will be taken into account. The inference engine will not sense whether the result arrived in time and was used or had been discarded.

If, however, there is a need to interrupt a lenghty reasoning process in order to perform an urgent reasoning task with high priority, then additional mechanisms, which are implemented in the real-time inference engine, are needed. In order to understand the problems of *interrupting reasoning*, we need to remember, that both forward and backward chaining writes marks and/or temporary values into the knowledge base while processing a reasoning task. Moreover, the elements of a knowledge base are highly related, therefore theoretically the whole knowledge base should be locked for the entire reasoning process unless it is possible to partition it and that has been done. Consequently a reasoning process can only be interrupted at a high cost, because it needs to store the whole knowledge base together with the status of the reasoning. In such a cases the inference engine and the knowledge base is reset instead, while the interrupted reasoning request is put back to its message queue and the next urgent request is processed.

The interrupt of the reasoning process above requires a *reset mechanism* to be implemented in the intelligent subsystem of a real-time expert system. The reset of a knowledge base when interrupting a reasoning process is performed by restoring the original value of the predicates in the case of forward chaining and by deleting the marks for backtrack in the case of backward chaining.

## 3. DATA EXCHANGE BETWEEN THE REAL-TIME AND THE INTELLIGENT SUBSYSTEMS

As we have seen before in this chapter, some data is already exchanged between the real-time and intelligent subsystems of a real-time system in synchronization and communication messages . The data content of a message, however, is limited to a time stamp, a sender identifier, a message identifier and to a few message parameters. But the reasoning process in an intelligent control system uses the value of measured data as facts which need to be transferred from their primary place in the real-time database to their destination in the part of the knowledge base that stores facts in the intelligent subsystem (see arrows on Fig. 6.1). Notice that the reasoning result only has a few data or knowledge items as its parameters. These items can most often be easily described by message parameters.

This section mainly deals with the type of data transfer above, which is directed from the data files in the real-time subsystem to the part of the knowledge base which stores facts.

A separate subsection is devoted to the special architecture used in the case of multiple parallel inference engines in a real-time expert system.

## 3.1 LOOSE DATA EXCHANGE

We know that the knowledge base of a rule-based expert system consists of two parts: facts and relationships. Facts are stored in the form of predicates, that is knowledge items with Boolean values. The notion of facts can be carried over to more general types of knowledge bases where they denote data-like knowledge elements changed by the external world and/or by reasoning process.

Facts are further classified into *root predicates*, the values of which only depend on the external environment of the expert system, that is, user input or measurements, and into *derived predicates*, which are intermediate or final results in the reasoning process.

As it has been mentioned several times before in this chapter, there is only a loose connection between the real-time and the intelligent subsystems in a real-time expert system. The messages that request reasoning are collected in message queues and are marked by a time stamp. *A corresponding set of signal values, which the root predicate values depend upon, ought to accompany the request* so that it can be performed. These signal values form a large data set, therefore they cannot simply be included in the message as its parameters. Consequently, these signal values can be obtained in two alternative ways.

1. include a pointer that points to a relevant *snapshot of the measured data file* in the message, and store the snapshot in the real-time database for later use by the intelligent subsystem,

2. update the relevant changes in the real-time database and *maintain a real-time mirror image* within the intelligent subsystem.

In both cases a *preprocessing* step is needed before any reasoning. This determines the logical values of the root predicates in the part of the knowledge base that contains facts from the signal values collected by the real-time subsystem.

In the first case, when snapshots are used for data transfer from the real-time subsystem to the intelligent one, the first sub-step of preprocessing a reasoning request is reading the snapshot indicated by the time stamp of the request from the real-time database. This is shown by a data connection arrow from the real-time data base to the knowledge base manager in Fig. 6.1.

It is important to note that in this case, a snapshot utility is needed in the intelligent subsystem which is dedicated to save a consistent set of measured data with the same time stamp. The snapshot utility is usually implemented as a service function of the real-time database manager that takes care of the appropriate resource management (i.e. locks the measured data file).

The second alternative requires the maintenance of a partial real-time image of the measured data that are needed for any possible reasoning request within the intelligent subsystem. This is usually implemented by constructing a separate high priority message queue from the primary processing real-time process to the inference engine or to a separate task within the real-time subsystem, which monitors measured data, and possibly performs preprocessing. When any of the signals needed by the intelligent subsystem changes, the primary processing task sends a message that contains the signal (measure data) identifier, its new value and status and a time stamp. These data are then fed into the mirror image of the measured data file within the intelligent subsystem.

Observe that in this case, the consistency of the mirror image of the measured data file is not automatically maintained. There can be cases when the mirror image does not contain the necessary signal values with the same time stamp - as the reasoning request would require. This is because there are various message queues with different priorities. A single queue from the real-time subsystem to the intelligent one would solve the problem at the price of loosing priority handling. This issue ought to be taken into account when designing of a real-time expert system.

As we have seen before, there can be cases where the need for a measured value at a given time is recognized during the reasoning process. The real-time archive in the real-time subsystem does contain this "past" information, usually in the form of signal change event messages. Therefore a special *signal value request message queue* is needed between the inference engine and the event handling or archive process in the real-time subsystem. The inference engine will normally wait for the signal value it requested (i.e. this is a *message exchange type connection*), thus this process-process connection is an exception to the loose data transfer concept applied in real-time expert systems between their real-time and intelligent subsystems.

With the tools and techniques above for data exchange between the real-time and intelligent subsystems, it is relatively easy to implement the *reset function of the reasoning process.* If a process in the real-time subsystem requires the interruption of the current reasoning process in order to process an urgent reasoning request, then the following sub-steps have to be carried out.

1. store the interrupted reasoning request message in its original queue by putting it back to be the first message to be processed,
2. erase the fact base and marks, that is fill the values with *nil* or *unknown*,
3. perform preprocessing for the new urgent reasoning request message in the usual way.

Finally, it is important to notice, that the *time stamp of the reasoning result* is derived directly from the time stamp of the reasoning request using the following methods.

- in the case of *reasoning for diagnosis*
  during backward chaining the time stamp of the result will usually be identical to that of the reasoning request unless the dynamics of the system and/or time series of measured data are taken into account. In the latter case the time stamp is the time when the cause of the fault or malfunction occurred.
- in the case of *reasoning for prediction*
  when forward chaining is applied, the time stamp is created during the prediction phase depending on the kind of the request, it will be the time when the derived or unwanted event occurred.

## 3.2 THE BLACKBOARD ARCHITECTURE

Until now we have assumed, that a single inference engine is present in the intelligent subsystem of the real-time system, which performs all the necessary reasoning requested by the real-time control system. There are cases, however, when more than one inference engines, possibly with different types of knowledge bases and/or reasoning capabilities, are available. These inference engines may be placed in various computers or processors of a distributed hardware architecture or may be implemented as separate processes within a multitasking environment. It may be necessary that these inference engines have a joint knowledge base that contains all or part of the knowledge they share. *The blackboard architecture is commonly used to manage knowledge bases shared by more than one inference engine* [50], [51], [52].

It is important to note that instead of the separation and loose communication introduced before, the cooperation between the real-time processes and the inference engine(s) in a real-time expert system can also be implemented using a blackboard for all the data and knowledge these active elements share. The reason why this is rarely the case will be explained later in this subsection when the consistency and the management of the blackboard is discussed.

The *basic idea* behind having a shared knowledge base for the inference engines working together on solving related tasks is very simple. Imagine several scientists trying to solve a problem, which is too difficult for anyone to solve on their own. They do not speak to each other, that is, direct communication is forbidden to avoid noise and chaos in the room. Alternatively, they have a large common blackboard, and each of them has got chalk to write with and a sponge to erase anything from the blackboard. They can all see the entire blackboard and modify its contents when and how they wish. The solution of the problem then evolves gradually on the blackboard as a result of their joint and co-operative activity. There is no boss, that is, no central control of any kind, the solution process is completely democratic.

The communication between any scientist and the others is restricted to writing and reading to and from the blackboard in a "broadcast" manner: one speaks to everyone "whom it may concern". If anyone notices a knowledge item s/he can contribute to, s/he elaborates on it immediately. This way, the scientists work in parallel in a knowledge-driven (data-driven) way.

This method is used for co-operating inference engines that are associated with the scientists. Their common evolving knowledge base is then called the blackboard. *The parallel, knowledge-driven operation and the democratic use of the blackboard is inherent in the concept.*

The software architecture of a multitasking real-time expert system where the inference engines operate in a parallel way sharing a common knowledge base is shown in Fig. 6.2.

The inference engines are denoted by rectangles, the blackboard-type knowledge base is the shadowed rectangle and the read-write data con-

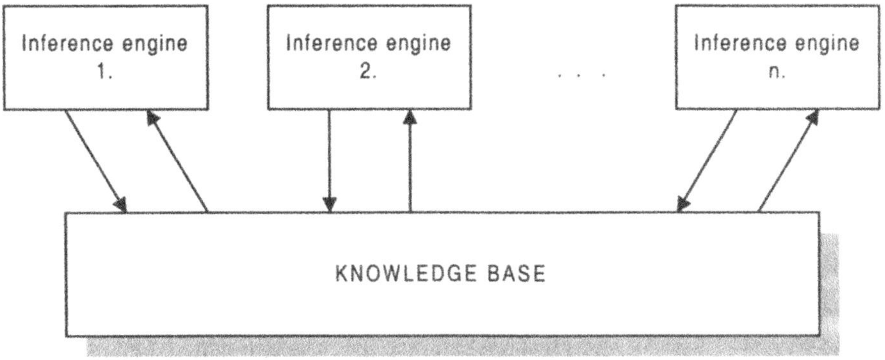


*Figure 6.2.* The blackboard architecture

nections between the active and passive elements are shown by arrows. It is important to notice and remember that *there is no direct communication and synchronization between the inference engines*, that is, between the active elements.

*The management of the blackboard* is a key issue in a blackboard architecture, because there is no central control of any kind to coordinate the parallel activity of the inference engines. A traditional knowledge base manager is not suitable for this purpose because its locks the entire knowledge base for the complete duration of a reasoning process, which is against the basic philosophy of the blackboard architecture.

We may apply a sophisticated extended database manager of a relational type to act as a blackboard knowledge base manager. There is a *tradeoff, however, between its efficiency and its consistency management capabilities*. If there are highly related knowledge elements and/or a non-decomposable blackboard knowledge base, the blackboard manager will either be inefficient because it is taking care of all the consequences of a change on the blackboard to ensure its consistency, or the blackboard will be inconsistent from time-to-time. This means that *the management of the blackboard determines how its consistency can be maintained.*

# 4. SOFTWARE ENGINEERING OF REAL-TIME EXPERT SYSTEMS

The basic approach, principles and methods of software engineering [53] does apply to real-time expert systems with some slight extensions and special features, which are consequences of both the real-time and the intelligent nature of this type of software systems. This section summarizes the basics of software engineering in the context of real-time expert systems [54], mainly focusing on the relevant special approaches, tools and techniques.

## 4.1 THE SOFTWARE LIFECYCLE OF REAL-TIME EXPERT SYSTEMS

The software lifecycle of a real-time expert system [55] is similar to that of other software systems. The schematic diagram in Fig. 6.3 indicates the stages in the software lifecycle of an intelligent control system that is a real-time expert system.

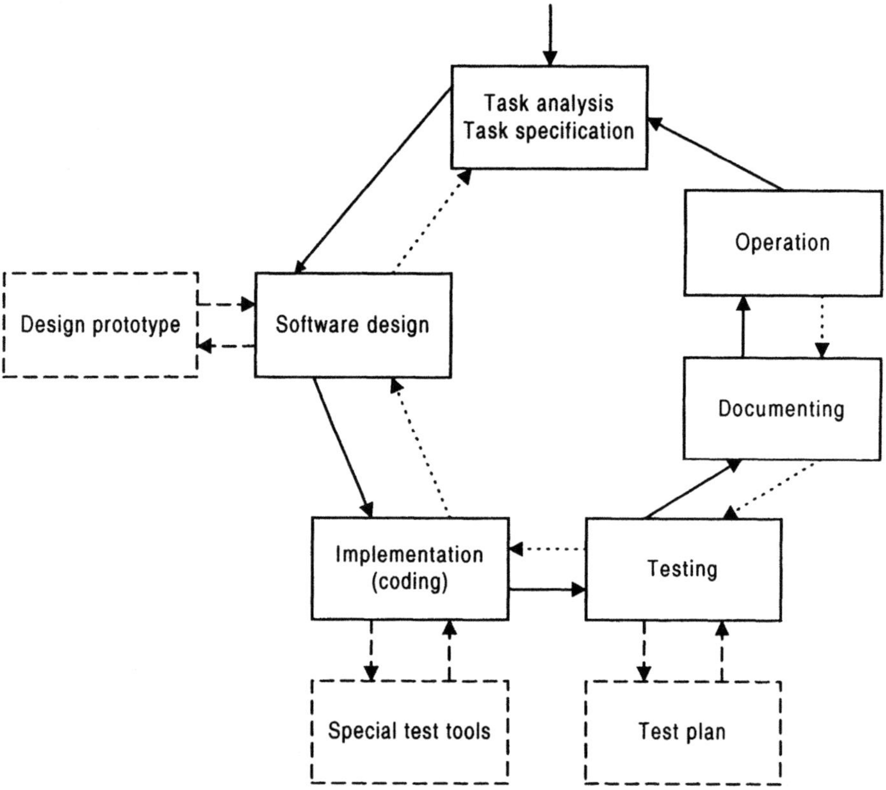


*Figure 6.3.* **The software lifecycle of intelligent control systems**

The common basic stages are shown in bold rectangles. These are extended by special sub-stages that reflect the needs of a real-time expert system and are denoted by dashed rectangles. The main flow of information during the "evolution" of a software system is indicated by bold arrows.

The main stages of a software lifecycle are the following.

1. *Task analysis and task specification*
   The first step in creating software is a decision on its construction. This is then followed by a "Task specification" which gives the problem statement and requirements for operation and implementation. Task specification is usually part of a contract when the implementation of the software is given to a software firm (department, company etc.). The task specification should be written by and for the future user of the software system, using an everyday language.

   Task analysis thereafter looks at the main consequences of task specification to determine the necessary resources and software tools needed for implementation. This way, task analysis prepares the ground for software design, which is the next stage. It is again important that this document should be a result of a joint effort on the part of the users and the implementers, including the knowledge engineer in the case of real-time expert systems. Still, it should focus on the user's viewpoint.

2. *Software design*
   Software design is a major document that describes the software in such detail that it can almost automatically be coded, even by individual technicians. Software design uses the terminology of software engineering, it can only be understood with background knowledge in computers and computing. This should include fundamentals of knowledge representation and reasoning in knowledge-based systems, such as real-time expert systems.

   There are computer-aided software engineering (CASE) tools available to be used in the lifecycle of a software, starting with software design stage in the case of common, i.e. non-intelligent software systems. Unfortunately, they cannot efficiently cope with the special features of real-time expert systems because of the great variety and experimental nature of the concepts, tools and techniques there.

   Instead, *individually made tools, the so called design prototypes are used to support software design and to test design alternatives in the software engineering of real-time expert systems.*

3. *Implementation (coding)*
   Coding is a relatively mechanical technical step in the implementation of a conventional software system. Part of coding can automatically

be done by the CASE tools mentioned before, using the formal software specification. The database is also constructed in this step, and the necessary data transfer is also carried out.

The implementation (coding) of a real-time expert system is far more complicated and involves many more creative elements. This is partly because *the knowledge needs to be elicited and validated to fill the relationship part and the factual part of the knowledge base* in this stage. *Expert system shells offer advanced tools and pre-fabricated elements to implement specific expert systems*, provided one manages to find an expert system shell that fits the purpose.

Besides the elements of the final software, special testing tools for testing and monitoring purposes are usually also coded here.

4. *Testing*
The main purpose of testing is to check if the completed software meets the criteria laid out in task specification (stage 1). Besides of user-oriented testing, all the algorithms and data elements need to be thoroughly verified and validated in all possible circumstances to ensure that the system operates properly. *Exhaustive testing is clearly not possible for*

   - neither *the intelligent subsystem because it is computationally hard to verify and validate the reasoning and the knowledge,*
   - nor *the real-time subsystem because of the high number of possible signal values and their timing combination that causes different real-time circumstances.*

   Therefore a special *test plan* needs to be set up well in advance to ensure the proper partial testing of the most important functions and the testing of the entire system under the most frequently occurring circumstances.

5. *Documenting*
Documenting is a standard but not very popular stage in a software lifecycle. There is nothing special in the documentation of a real-time expert system, it will simply reflect the strong relationships between the elements and therefore will be strongly inter-related.

   Advanced tools, such as CASE tools and expert system shells support documentation, sometimes even self-documentation is possible (compare with the service debugging tools of an expert system shell in section 3. of Chapter 5).

6. *Operation*
Before a software system is put into operation, users and operating

personnel are trained. Training includes the education of proper software maintenance. This also applies to real-time expert systems.

One of the inherent characteristics of any software lifecycle including this one in Fig. 6.3 is its repetitive or bidirected nature shown by dotted arrows. If the completion of a step results in a non-satisfactory product, then there is a need to step back to the previous stage to investigate and possibly repeat or partially repeat it to correct the problems. If this cannot be done by repeating the previous step, then one should step back again from here to an earlier step to find the cause. In the worst case one may end at the first stage, "Task analysis and Task specification" and correct the original problem statement.

## 4.2 THE SPECIAL STEPS AND TOOLS IN DEVELOPING AND IMPLEMENTING A REAL-TIME EXPERT SYSTEM

The main stages of the software lifecycle of a real-time system have been introduced, in the subsection above. Here we focus on the special sub-stages, which serve the proper development of a real-time software system.

1. *Design prototype*
   The aim of a design prototype in designing a real-time software system is to test and demonstrate the operation of a partial solution, knowledge representation and/or reasoning technique. Usually only a part of the knowledge base is used and only a few key functions are implemented, therefore a *design prototype is a downsized partial version of the final system.* Sometimes there are a few different design prototypes used for the same system to test various aspects of the final version.

   There is, however, a major danger in applying a design prototype and carrying over a positive result to the final full-size system. To understand this we need to remember, that the reasoning, verification and validation of a knowledge base are all computationally hard, therefore the number of computational steps may grow exponentially (non-polynomially) with the size of the knowledge base. This implies that *acceptable response times on a design prototype do not carry over to the final full-size version.*

2. *Special testing tools*
   Expert system shells usually offer special advanced testing tools to test the stand-alone functions of the intelligent subsystem (see debugging tools of an expert system shell in section 3. of Chapter 5),

thus special testing tools are needed to test the interface and the operation of the real-time expert system under different time-varying real-time conditions.

The need for special testing tools is explained by the fact that it is quite difficult to test any real-time system properly, and this situation is even worse if intelligent elements are present.

Special test tools usually include

- programmable test signal generators that imitate the behaviour of the real plant of the real-time subsystem,
- interface monitors that monitor the status of message queues
- a test archive process, which logs every change in the intelligent subsystem, for example using special event messages.

3. *Testing plan*
As it has been explained before, an exhaustive test is not feasible for real-time expert system, therefore, a testing plan prepared well in advance is a must. The following situations in a test plan should be treated with special care:

   - normal operation with various extreme loads (extreme high and extreme low) and their transients,
   - abnormal and faulty operation modes, such as time-outs, missing or faulty elements, non-available resources, reset requests, missing data, corrupted data etc. and their transients such as start-up, shut-down, going to degradated mode, restoring normal behaviour from degradated mode etc.
   - conflicting data and/or knowledge elements
   - test mode operation (with the special testing tools operating on-line)

Finally, we would like to emphasize again that the software engineering of real-time expert systems and especially the development and use of the special tools above are far from being matured. The creativity and knowledge of the software engineer who is specialized in this field are more than necessary in all cases.

# Chapter 7

# QUALITATIVE REASONING

The aim of qualitative modelling is to describe partially known systems for control and diagnostic purposes [56], [57]. The known elements form the structure of the model which is equipped with interval valued or symbolic elements in order to describe the unknown part. The presence of interval valued or symbolic elements in a qualitative model calls for the application of AI techniques, namely special reasoning to perform prediction or decision making for diagnosis or control. Therefore qualitative reasoning [58], the subject of this chapter, is applied as a special technique in intelligent control systems.

The following qualitative reasoning methods [59], [58] are described and compared in this chapter.

- Qualitative simulation [60] in section 2.
- Qualitative physics [61] in section 3.
- Signed Directed Graph models [62] in section 4.

All three methods above use sign and/or interval calculus for qualitative reasoning which will be discussed in a separate section first.

The origin of any qualitative model of the types above is the nonlinear state-space model of lumped or concentrated parameter system models the general form of which is the following:

$$\frac{d\underline{x}}{dt} = \mathbf{f}(\underline{x}, \underline{u}) \quad , \quad \underline{x}(0) = \underline{x}_0 \tag{7.1}$$

$$\underline{y} = \mathbf{g}(\underline{x}, \underline{u}) \tag{7.2}$$

We may linearize it around a steady-state point $(\underline{x}^*, \underline{u}^*)$ to obtain the linear(ized) version of the above state-space model in the form:

$$\frac{d\underline{x}}{dt} = \mathbf{A}\underline{x} + \mathbf{B}\underline{u} \quad , \quad \underline{x}(0) = \underline{x}_0 \tag{7.3}$$

$$\underline{y} = \mathbf{C}\underline{x} + \mathbf{D}\underline{u} \tag{7.4}$$

where the constant matrices $(\mathbf{A}, \mathbf{B}, \mathbf{C}, \mathbf{D})$ are the parameters of the linear time-invariant model above.

As we shall see later in this chapter, both qualitative simulation and qualitative physics uses the full nonlinear state-space model, while the signed directed graph models correspond to the linear(ized) version.

# 1. SIGN AND INTERVAL CALCULUS

In comparison with the "traditional" engineering models, qualitative, logical and artificial intelligence (AI) models have a special common property: *the range space of variables and expressions in these models is interval valued.* This means that we specify an interval $[a_{\ell_t}, a_{u_t}]$ for a variable $a$ at any given time within which the value of the variable lies. Thereafter every value of the variable within the specified interval is regarded to be the same in a qualitative sense because all of these values have the same qualitative value. This means that the value of the variable $a$ can be described by a finite set of non-intersecting intervals covering the whole range space of the variable.

In the most *general qualitative case* these intervals are real intervals with fixed or free endpoints. The so called *universe* of the range space of the interval-valued variables $U_{\mathcal{I}}$ in this case is

$$U_{\mathcal{I}} = \{[a_\ell, a_u] \mid a_\ell, \; a_u \in \mathcal{R}, \; a_\ell \leq a_u\} \tag{7.5}$$

Observe that the above universe is generated by a finite or infinite number of points

$$L_{\mathcal{I}} = \{a_i \mid a_i \leq a_{i+1} \; , \; i \in I \subseteq \mathcal{N}\}$$

There are models with *sign-valued* variables. Here the variables may have the qualitative value "+" when their value is strictly positive, the qualitative values "−" or "0" when the real value is strictly negative or exactly zero. If the sign of the value of the variable is not known then we assign to it an "unknown" sign value denoted by "?". Note that the sign value "?" can be regarded as a union of the three other sign values above. It means that if the value of a sign valued variable is unknown then it may either be positive "+", zero "0" or negative "−". The corresponding universe $U_S$ for the sign valued variables is in the form

$$U_S = \{ + \; , \; - \; , 0 \; ; \; ? \} \; , \; ? = + \cup 0 \cup - \tag{7.6}$$

It is important to note that the sign universe is a special case of the interval universe generated by the points

$$L_S = \{ \; a_1 = -\infty \;, \; a_2 = 0 \;, \; a_3 = \infty \}$$

with the intervals

$$U_S \; = \; \{ \; (a_1, a_2) \;, \; [a_2, a_2] \;, \; (a_2, a_3) \; \}$$

Finally, *logical models* operate on logical variables. Logical variables may have the value "**true**" and "**false**" according to the traditional mathematical logic. If we consider time varying or measured logical variables then their value may also be "**unknown**". Again, note that the logical value "**unknown**" is the union of "**true**" and "**false**". The universe for logical variables is then:

$$U_{\mathcal{L}} \; = \; \{ \; \textbf{true} \;, \; \textbf{false} \;; \; \textbf{unknown} \; \} \tag{7.7}$$

## 1.1 SIGN ALGEBRA

Sign algebra is applied for variables and expressions with sign values, where their range space is the so called *sign universe* defined in Eq. (7.6).

We can consider the sign universe as an extension of the logical values ("**true**"," **false**", "**unknown**") forming the logical universe in Eq. (7.7) with "**true**" being +, "**false**" being − extended by 0. Thus we can define the usual arithmetic operations on sign-valued variables with a help of operation tables. The operation table of the sign addition ($\oplus_S$) and that of the sign multiplication ($\otimes_S$) is given below.

The operation table for the sign substraction and division can be defined analogously. Note that the operation tables of functions and other operators, such as *sin*, *exp*, etc., can be generated from the Taylor expansion of the functions, using the operation tables of the elementary algebraic operations.

*Table 7.1.* Operation table of the sign addition

| $a \oplus_S b$ | + | 0 | − | ? |
|---|---|---|---|---|
| + | + | + | ? | ? |
| 0 | + | 0 | − | ? |
| − | ? | − | − | ? |
| ? | ? | ? | ? | ? |

*Table 7.2.* Operation table of the sign multiplication

| $a \otimes_S b$ | + | 0 | − | ? |
|---|---|---|---|---|
| + | + | 0 | − | ? |
| 0 | 0 | 0 | 0 | **0** |
| − | − | 0 | + | ? |
| ? | ? | **0** | ? | ? |

It is important to note that sign algebra has the following important properties.

1. *growing uncertainty with additions*, which is seen from the table of sign addition as the result of "$+ \oplus_S -$" being unknown "?",
2. the usual algebraic properties of addition and multiplication, i.e. commutativity, associativity and distributivity.

## 1.2 INTERVAL ALGEBRAS

The basic operations defined on intervals [63], [64] with fixed endpoints exhibit some unusual properties which is the consequence of their so called *set-type definitions.* If we consider the universe of intervals with fixed endpoints in Eq. (7.5) then the basic algebraic operations $\oplus_I$ and $\otimes_I$ can be defined as follows.

**Definition 7.1.** *The sum (or product) of two intervals $\mathcal{I}_1 = [a_{1\ell}, a_{1u}]$ and $\mathcal{I}_2 = [a_{2\ell}, a_{2u}]$ from $U_\mathcal{I}$ is the smallest interval from $U_\mathcal{I}$ which covers the interval*

$$\mathcal{I}^* = \{\ b = a_1 \ \mathbf{op}\ a_2 \mid a_1 \in \mathcal{I}_1\ ,\ a_2 \in \mathcal{I}_2\} \tag{7.8}$$

*where* **op** *is the usual sum or product on real numbers respectively.*

In the case of monotonic operations with respect to their arguments, like sum and product we can compute the result of the above set type definitions using only the endpoints of the two intervals in the following way.

$$\begin{aligned} E_{op} = \{\ & e_{\ell\ell} = a_{1\ell}\ \mathbf{op}\ a_{2\ell}\ ,\ e_{\ell u} = a_{1\ell}\ \mathbf{op}\ a_{2u}\ , \\ & e_{u\ell} = a_{1u}\ \mathbf{op}\ a_{2\ell}\ ,\ e_{uu} = a_{1u}\ \mathbf{op}\ a_{2u}\ \} \end{aligned} \tag{7.9}$$

$$\mathcal{I}^* \ = \ [\ min\ E_{op},\ max\ E_{op}\ ] \tag{7.10}$$

where *min* $E_{op}$ is the smallest element and *max* $E_{op}$ is the largest element in the set $E_{op}$ formed from the endpoints of the individual intervals. The endpoint type calculation above enables us to perform interval operations in polynomial time whereas the original set definition is computationally hard.

It is important to note that the resulting interval $\mathcal{I}^*$ is to be covered by adjacent intervals from the original interval universe $U_{\mathcal{I}}$, which is usually a conservative operation. This means that the covering interval is usually larger than the original $\mathcal{I}^*$. *This fact results in a natural increase in the uncertainty of all kinds of operations over intervals with fixed endpoints.*

In order to illustrate the use and possible problems of interval operation the simplest case, the so called *order of magnitude interval algebra* is considered. Here the interval universe is generated by five points which are put into the real line in a symmetric way:

$$L_{\mathcal{OM}} = \{ \; a_1 = -\infty \; , \; a_2 = -A \; , \; a_3 = 0 \; , \; a_3 = A \; , \; a_4 = \infty \} \qquad (7.11)$$

where $A > 0$ is a constant. With the points above the following elementary (that is non-divisible) intervals are formed in the order of magnitude interval universe:

$$U_{\mathcal{OM}} \;=\; \{ \; LN \; , \; SN \; , \; 0 \; , \; SP \; , \; LP \; \} \qquad (7.12)$$

with

$$LN = [-\infty, -A), \;\; SN = [-A, 0), \;\; 0 = [0, 0], \;\; SP = (0, A], \;\; LP = (A, \infty]$$

It can be seen that the above universe is just a little bit more fine that the sign universe in Eq. (7.6).

Like logical and sign operations, interval operations are also defined using operation tables. The table 7.3 below shows an example of this, being the operation table of addition over the order of magnitude interval universe. If the result of a particular operation can only be covered by more than one adjoint elementary interval from the universe (7.12), then a pseudo-interval showing the endpoint-intervals of the covering interval is shown in the table, for example

$$[SP, LP] = (0, \infty] \quad or \quad [LN, LP] = [-\infty, \infty]$$

It is obvious from the table that the interval addition over the order of magnitude interval universe is commutative because the table is symmetric. The growing uncertainty property is also clear if we compare the width of the original and the resulting intervals, for example

$$LP \; \oplus_{OM} \; LN \;=\; [LN, LP]$$

*Table 7.3.* Operation table of order of magnitude interval addition

| $a \oplus_{OM} b$ | $LN$ | $SN$ | $0$ | $SP$ | $LP$ |
|---|---|---|---|---|---|
| $LN$ | $LN$ | $LN$ | $LN$ | $[LN, SN]$ | $[LN, LP]$ |
| $SN$ | $LN$ | $[LN, SN]$ | $SN$ | $[SN, SP]$ | $[SP, LP]$ |
| $0$ | $LN$ | $SN$ | $0$ | $SP$ | $LP$ |
| $SP$ | $[LN, SN]$ | $[SN, SP]$ | $SP$ | $[SP, LP]$ | $LP$ |
| $LP$ | $[LN, LP]$ | $[SP, LP]$ | $LP$ | $LP$ | $LP$ |

It is important to note that the any of the interval algebras has the following important properties.

1. *growing uncertainty with additions and also with multiplication*
   This causes all of the other elementary or composite algebraic operations to possess a growing uncertainty property.

2. *the usual algebraic properties of addition and multiplication, i.e. commutativity and associativity, but* **not** *distributivity*
   This means that the evaluation of algebraically equivalent expressions does not necessarily give the same result. It can be shown that the form with the minimum number of addition gives the best, that is, the narrowest result.

# 2. QUALITATIVE SIMULATION

Qualitative simulation operates on the finest qualitative models, the so called constraint type qualitative differential equation models (QDEs) [60], [65]. The solution of a constraint type QDE is generated by a combinatorial algorithm, by qualitative simulation.

## 2.1 CONSTRAINT TYPE QUALITATIVE DIFFERENTIAL EQUATIONS

The syntax of constraint type QDEs is exactly the same as that of the usual nonlinear state-space models in Eqs. (7.1)-(7.2). This means that we can formally use any model in the form of ordinary differential and algebraic equations as a constraint type QDE model.

When compared to an usual nonlinear state-space model, the essential difference lies in the range space of the variables and the parameters of a constraint type QDE. The range space in the former model is the set of fixed endpoint intervals which calls for the application of interval arithmetics in the model equations. Moreover, qualitative functions can also be part of a constraint type QDE model.

In order to construct a constraint type QDE model of a system we start from its nonlinear state-space model equations (7.1)-(7.2). The ingredients of constraint type QDEs, that is the essential elements to be defined when constructing a model are as follows.

1. *Qualitative counterparts of the variables*
   Any time-dependent signal in the model equations is regarded as a variable. For any variable $q(t)$ at any time $t$ in the original nonlinear state-space model we associate a qualitative counterpart $Q(t)$ by first defining the range space of $Q(t)$ using the so called *landmark set* of $q(t)$, which is an ordered set of landmarks $\ell_{q_j}$:

$$L_q = \{\ell_{q_1}, \ldots, \ell_{q_n}\} \quad , \quad \ell_{q_{i-1}} < \ell_{q_i} \tag{7.13}$$

   It is important to note that any of the landmarks may have a numerical or a symbolic value as well. With the landmark set above *the value of the qualitative variable $Q(t)$ at any time $t$ is the following ordered pair:*

$$Q(t) =< Q_{mag}, Q_{dir} > \tag{7.14}$$

   where $Q_{mag}$ is the magnitude of the variable, which can be any of the landmarks from the landmark set (7.13) or any open interval formed by landmarks as endpoints, that is:

$$Q_{mag} = \begin{cases} \ell_{q_i} & \\ (\ell_{q_i}, \ell_{q_j}) & , \ j > i \end{cases} \tag{7.15}$$

   The second element $Q_{dir}$ in the value of a qualitative variable is its *direction of change*, which may have three different distinct values "increasing", "decreasing" and "steady" as follows:

$$Q_{dir} \in \{ \ inc, \ dec, \ std \ \} \tag{7.16}$$

   with

$$Q_{dir} = \begin{cases} inc, & if \ \frac{dq(t)}{dt} > 0 \\ dec, & if \ \frac{dq(t)}{dt} < 0 \\ std, & if \ \frac{dq(t)}{dt} = 0 \end{cases} \tag{7.17}$$

2. *Qualitative counterparts of the parameters*
   The parameters are transformed to qualitative pseudo-parameters by forming qualitative variables with an identically "*std*" (steady) qualitative direction of change. This way a parameter $k$ in an ordinary

state-space model will be transformed to a qualitative parameter $K$ with

$$K = < k, std > \quad or \quad K = < (k_1, k_2), std >$$

depending on the landmark set applied.

3. *Qualitative functions*
One of the distinctive characteristics of constraint type QDEs is that they may contain qualitative functions, that is, sets of given functions as their pseudo-parameters. It is usually required that any member of the set have prescribed properties, such as monotonicity.

There are two possible ways to specify a qualitative function: by giving corresponding values or by describing envelope functions. The two specification methods are illustrated here with the example of a single variable real-valued qualitative function $F_Q(.)$, the qualitative counterpart of a real-valued univariate function $f(.)$.

- *corresponding values*
In order to specify the set of ordinary real-valued functions that belong to the set we specify data points, that is, $(x, f(x))$ pairs through which every member function should go. Thus the corresponding values form a set

$$C_V = \{ \ (x_1, y_1), \ \ldots, \ (x_n, y_n) \}$$

Moreover, all the member functions should be monotonous. This means that a real-valued function $f$ is a member function in $F_Q$ $(f \in F_Q)$ specified by the corresponding value set $C_V$ if

$$f(x_i) = y_i \quad , \quad i = 1, \ldots, n$$

and $f$ is monotonous.
Figure 7.1 shows an example when the qualitative function is given by three corresponding values denoted by bold dots. Two possible member functions are also shown, both of them are monotonously increasing.

- *envelopes*
We may specify two envelope functions ($f_\ell$ and $f_u$) which encapsulate the set of member functions in a qualitative function $F_Q$. Here again it is required that both the envelope and the member functions be monotonous. Formally speaking any real-valued function $f$ is a member of the set $F_Q$ $(f \in F_Q)$ specified by the envelope functions $f_\ell$ and $f_u$ if

$$f_\ell(x) \leq f(x) \leq f_u(x) \quad \forall x \quad .$$

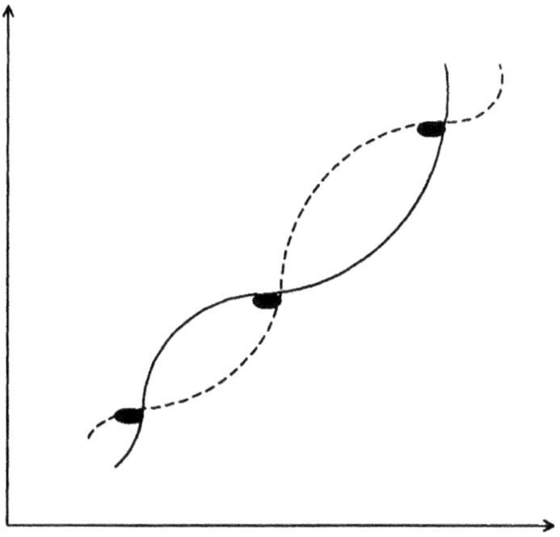

*Figure 7.1.* Qualitative functions given by corresponding values

Figure 7.2 shows a simple example with two monotonously increasing envelope functions and two possible member functions.

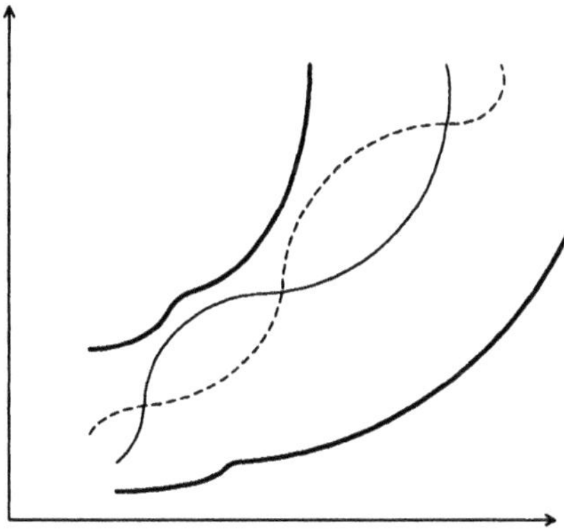

*Figure 7.2.* Qualitative functions given by envelope functions

4. *Qualitative time*
   The notion of time is also extended in qualitative simulation. Time is measured using the so called *qualitative time set consisting of distinguished time points* $t_i$. Any point in real time when any of the qualitative variables $Q^{(j)}(t)$ changes its qualitative value generates a

distinguished time point $t_i$.

$$\begin{aligned} T \quad = \quad & \{\ t_i \mid t_i < t_{i+1}\ ,\ Q^{(j)}_{mag}(t_i) := \ell^{(j)}_{q_k} \\ \textit{and} \quad & Q^{(j)}_{dir}(t_i) := inc \ \textit{or} \ Q^{(j)}_{dir}(t_i) := dec \\ \textit{or} \quad & Q^{(j)}_{dir}(t_i) := std \ \ for \ any \ j\} \end{aligned} \tag{7.18}$$

In other words, events caused by any qualitative value change in the system generate a discrete distinguished time point.

5. *The qualitative behaviour of a system*
   The qualitative behaviour of a system is described using its so called qualitative state. The qualitative state of a system is simply the set of the qualitative values of all of its qualitative state variables. Note that not only the usual state variables of the system are considered here but all of its input and output variables as well. Let us denote the qualitative state of a system at a given distinguished time point $t_0$ by $S(t_0)$ and between two adjacent distinguished time points $(t_0, t_1)$ by $S(t_0, t_1)$.

   The *qualitative behaviour of a system between two distinguished time points $t_0$ and $t_k$* is then *a sequence of qualitative states* in the form of:

$$D(t_0, t_k) = (\ S(t_0),\ S(t_0, t_1),\ S(t_1),\ \ldots,\ S(t_{k-1}, t_k),\ S(t_k)\ ) \tag{7.19}$$

6. *The constraint type QDE model*
   The constraint type QDE model has the same algebraic form as the usual model with its ordinary differential and algebraic equations but its variables and parameters are qualitative. It may also contain qualitative functions with an appropriate specification. Therefore the algebraic manipulations in the model equations are understood as being qualitative (interval) manipulations.

The above concepts are illustrated with the example of the batch water heater (coffee machine) introduced in Appendix B.

**EXAMPLE 7.1** The constraint type QDE model of the batch water heater in Appendix B

The QDE model equations are derived from the model equations (B.1)-(B.2) through the following steps:

1. *Qualitative variables*
   The following time-dependent signals are considered:

   $h$ level in the tank with the landmark set

$$L_h = \{0, h_l, h_u, h_m\} \tag{7.20}$$

   where $h_l$ is low, $h_u$ is full and $h_m$ is the maximum level.

   $T$ temperature with the landmark set

$$L_T = \{0^oC, T_I, T_{op}, 100^oC\} \tag{7.21}$$

   where $T_I$ is the inlet and $T_{op}$ is the operating (ready) temperature.

   $\kappa$, $\eta_I$ and $\eta_O$ switches with the joint landmark set

$$L_{sw} = \{0, 1\} \tag{7.22}$$

   where 0 corresponds to the closed and 1 to the open status.

   All the other symbols are considered to be parameters described by suitable qualitative constants.

2. *Constraint type QDEs*
   are formally the same as in Eq. (B.1)-(B.2):

$$\frac{dh}{dt} = \frac{v}{A}\eta_I - \frac{v}{A}\eta_O \tag{7.23}$$

$$\frac{dT}{dt} = \frac{v}{Ah}(T_I - T)\eta_I + \frac{H}{c_p \rho h}\kappa \tag{7.24}$$

3. *Qualitative state*
   The qualitative state of the system $S$ consists of the following qualitative variables:

$$S(.) = \{\ h(.),\ T(.),\ \kappa(.),\ \eta_I(.),\ \eta_O(.)\ \} \tag{7.25}$$

## 2.2 THE SOLUTION OF QDES: THE QUALITATIVE SIMULATION ALGORITHM

There are in principle two possible ways of solving the constraint type QDEs, the model equations of qualitative simulation.

1. Numerical solution using interval arithmetic methods

2. Algorithmic solution by qualitative simulation

This section describes the second way of solution. Note that the algorithmic solution is equivalent to a numerical Euler method for solving ordinary differential equations in the limit [66].

*Qualitative simulation is an algorithmic method which uses the constraint type QDEs and the initial state of the system as knowledge items and generates the set of possible qualitative behaviours of the system.* The algorithm of qualitative simulation is called the QSIM algorithm.

The elements and main steps of the QSIM algorithm are given below.

### 2.2.1 INITIAL DATA FOR THE QUALITATIVE SIMULATION

In order to perform qualitative simulation to generate the solution of a constraint type QDE model, one needs to give the following as input data to the algorithm.

1. A symbol set for the qualitative variables of the system

$$\mathbf{X} = \{ X_1, X_2, \ldots, X_n \}$$

2. Constraint type QDE model equations over set $\mathbf{X}$

3. A landmark set for every variable $X_i \in \mathbf{X}$

4. The domain of the model
   The landmark set specifies a finite domain for every variable, which determines the domain of the overall model. In some cases we may give several sets of constraint type QDEs and specify subdomain of the overall domain over which a specific set is applicable. When reaching the boundary of a subdomain one should switch to another model.

5. The initial state of the system

The initial state $S(t_0)$ of the system is the value of all of its qualitative variables in the initial distinguished time point $t_0$. The qualitative magnitude of the variables is given in problem specification. The qualitative direction is computed using the constraint type QDEs and evaluating their right hand sides with interval arithmetics to determine the sign of time derivatives. This step is called the ***augmentation of the initial state***.

The following simple example shows how the augmentation of the initial state is done.

---

**EXAMPLE 7.2** (Example 7.1 continued)
The augmentation of the initial qualitative state of the batch water heater in Appendix B

A possible initial state of the batch water heater is characterized by the following qualitative magnitude of the variables:

$$h_{mag}(t_0) = 0 \ , \ T_{mag}(t_0) = T_I$$

corresponding to an empty tank with the temperature equal to the inlet water temperature. Moreover, we fix the input variables to be

$$\kappa_{mag}(t_0) = 1 \ , \ \eta_{I,mag}(t_0) = 1 \ , \ \eta_{O,mag}(t_0) = 0$$

which means that we switch the heating and the inlet flow on and close the outlet. The qualitative direction of the input variables above is fixed to "*std*".

The constraints (7.23) and (7.24) are used to compute the qualitative direction of the variables $h$ and $T$. With all the parameters being positive constant we finally obtain:

$$h(t_0) =< 0, inc > \quad , \quad T(t_0) =< T_I, inc >$$

The qualitative state of the system at the distinguished initial time point $t_0$ is then

$$S(t_0) = \{< 0, inc >, \ < T_I, inc >, \ < 1, std >, \ < 1, std >, \ < 0, std >\} \tag{7.26}$$

---

### 2.2.2 STEPS OF THE SIMULATION ALGORITHM

The qualitative simulation algorithm is a special way of solving constraint type QDEs. It is based on a ***basic assumption*** on the variation of system variables in time: *it is assumed that both a system variable and its time derivative are continuous functions of time.* With this assumption one can reason on the next qualitative value of any variable with a given qualitative value by making use of the fact that there cannot be any jumps, neither in its qualitative magnitude nor in its qualitative direction of change.

Using the augmentation procedure described above, the qualitative simulation itself starts at the initial time $t_0$ as the first distinguished time point by computing the initial qualitative state

Thereafter the simulation is performed by successively generating and examining the distinguished time points. This implies that the steps of the simulation algorithm are twofold:

1. the system either moves from a distinguished time point $t_i$ to its succeeding open interval $(t_i, t_{i+1})$ by a so called *P-transition*,

2. or it terminates an open interval $(t_i, t_{i+1})$ by computing a new distinguished time point $t_{i+1}$ by a so called *I-transition*

As the simulation proceeds the algorithm performs a sequence of steps of this type:

$$(P\!-\!transition)\!-\!(I\!-\!transition)\!-\!(P\!-\!transition)\!-\!(I\!-\!transition)\ldots$$

In each step the following substeps are performed both in the case of *P-* and *I-transitions*.

Br1 *Examine each of the system variables separately for its possible change in qualitative value* and generate branches accordingly.

Br2 *For a given changing variable generate all the next possible qualitative values using the continuity assumption* and generate branches accordingly. The next possible qualitative values are given in a table separately for *P-transition* (see Table 7.4) and *I-transition* (see Table 7.5).

Bo *For a given changing variable evaluate the right-hand side of the constraint type QDEs with the new qualitative magnitudes to obtain the possible qualitative directions of all the variables. Use these qualitative directions to cut these branches that contradict to the model equations.*

Observe that the algorithm above is a "branch-and-bound" type algorithm where the branches are generated in substeps "Br1" and "Br2"

*Table 7.4.* The table of $P$-transitions

| $P$ | $Q(t_i)$ | $Q(t_i, t_{i+1})$ | remark |
|---|---|---|---|
| $P1$ | $< \ell_k, std >$ | $< \ell_k, std >$ | change elsewhere |
| $P2$ | $< \ell_k, std >$ | $< (\ell_k, \ell_{k+1}), inc >$ | start increasing |
| $P3$ | $< \ell_k, std >$ | $< (\ell_k, \ell_{k+1}), dec >$ | start decreasing |
| $P4$ | $< \ell_k, inc >$ | $< (\ell_k, \ell_{k+1}), inc >$ | landmark passed |
| $P5$ | $< \ell_k, dec >$ | $< (\ell_k, \ell_{k+1}), dec >$ | landmark passed |
| $P6$ | $< (\ell_k, \ell_{k+1}), inc >$ | $< (\ell_k, \ell_{k+1}), inc >$ | change elsewhere |
| $P7$ | $< (\ell_k, \ell_{k+1}), dec >$ | $< (\ell_k, \ell_{k+1}), dec >$ | change elsewhere |

*Table 7.5.* The table of $I$-transitions

| $I$ | $Q(t_i, t_{i+1})$ | $Q(t_{i+1})$ | remark |
|---|---|---|---|
| $I1$ | $< \ell_k, std >$ | $< \ell_k, std >$ | change elsewhere |
| $I2$ | $< (\ell_k, \ell_{k+1}), inc >$ | $< \ell_{k+1}, inc >$ | landmark reached |
| $I3$ | $< (\ell_k, \ell_{k+1}), inc >$ | $< \ell_{k+1}, std >$ | stop increasing at landmark |
| $I4$ | $< (\ell_k, \ell_{k+1}), inc >$ | $< (\ell_k, \ell_{k+1}), inc >$ | change elsewhere |
| $I5$ | $< (\ell_k, \ell_{k+1}), dec >$ | $< \ell_k, dec >$ | landmark reached |
| $I6$ | $< (\ell_k, \ell_{k+1}), dec >$ | $< \ell_k, std >$ | stop decreasing at landmark |
| $I7$ | $< (\ell_k, \ell_{k+1}), dec >$ | $< (\ell_k, \ell_{k+1}), dec >$ | change elsewhere |
| $I8$ | $< (\ell_k, \ell_{k+1}), inc >$ | $< \ell^*, std >$ | stop increasing before landmark |
| $I9$ | $< (\ell_k, \ell_{k+1}), dec >$ | $< \ell^*, std >$ | stop decreasing before landmark |

and the model equations are used for "constraining" these branches, that is for bounding in substep "Bo".

It can be seen from the transition tables 7.4 and 7.5 that the next state is not unique in the general case therefore the qualitative simulation algorithm is not polynomial in the number of qualitative time steps. Furthermore the branching substep "Br1" generates at least as many branches as many system variables we have therefore the algorithm is not polynomial in the number of system variables either.

### 2.2.3 SIMULATION RESULTS

The algorithm of qualitative simulation incrementally generates the set of all possible behaviours $D(t_0, t_k)$, $k = 1, 2, \ldots$ arranged in a so called behaviour tree.A vertex in a behaviour tree is a qualitative system state in either a distinguished time point $S(t_i)$ or in an open interval between two succeeding distinguished time points $S(t_i, t_{i+1})$. The root of the tree is the unique initial qualitative system state $S(t_0)$. There is a directed edge $(S(t_i), S(t_i, t_{i+1}))$ or $(S(t_i, t_{i+1}), S(t_{i+1}))$ in a tree if there is a transition, that moves the system from the initial state to the final state of the edge.

Branches occur in each of the $P - transition$ and $I - transition$ steps generated by the change of the qualitative system variables and by the non-uniqueness of the next qualitative states.

A possible behaviour $D(t_0, t_k)$ is then a directed path between $S(t_0)$ and a state corresponding to $t_k$ $S^{Pi_1 Ii_1 \ldots Pi_{k-1} Ii_{k-1}}(t_k)$, where $i_j$ is the index set that identifies the types of transitions for the variables in step $j$.

Figure 7.3 shows part of a behaviour tree when only one system variable is considered between the distinguished time points $t_i$ and $t_{i+1}$. It is also assumed that the model equations do not put any constraint to the qualitative behaviour, that is no branches are cut.

The simple example of the coffee machine will be used to illustrate the operation of the qualitative simulation algorithm.

---

**EXAMPLE 7.3** (Example 7.1 continued)
Generation of the qualitative behaviour of the batch water heater in Appendix B by qualitative simulation

Let us assume that we fix the value of the input variables to be

$$\kappa =< 1, std > \ , \ \eta_I =< 1, std > \ , \ \eta_O =< 0, std >$$

for the entire simulation. This means that we regard them as constants.

The augmentation of the initial qualitative state of the batch water heater is given in an earlier example, Example 7.2, but now we only need to consider the level and the temperature as state variables. Therefore the initial qualitative state of the system is a special case of Eq. (7.26):

$$S(t_0) = \{ \ < 0, inc > \ , \ < T_I, inc > \ \} \tag{7.27}$$

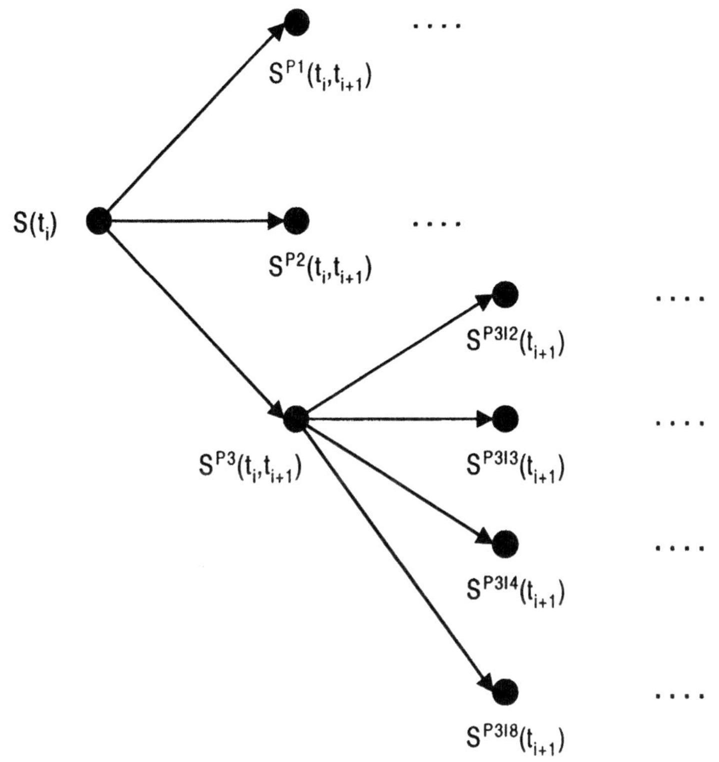


*Figure 7.3.* Part of a behaviour tree

*P-transition*
From the P-transition table 7.4 we find that the only possible transition is $P4$ for both of the state variables with the given initial state (7.27). Therefore *the next qualitative state* $S^{P4}(t_0, t_1)$ *is unique*:

$$S^{P4}(t_0, t_1) = \{ \ < (0, h_l), inc > \ , \ < (T_I, T_{op}), inc > \ \} \qquad (7.28)$$

*I-transition*
Now we examine the next possible values separately for the two variables from the I-transition table 7.5. There are four possibilities for each variable: $I2$, $I3$, $I4$ and $I8$.

Thereafter the constraints (7.23) and (7.24) are used to compute the qualitative direction of the variables $h$ and $T$. Eq. (7.23) implies that $h_{dir} = inc$ for the entire simulation, therefore only transitions $I2$ and $I4$ remain possible for $h$. The qualitative direction of the temperature is not constrained by Eq. (7.24) because the sign of the right hand side depends on the actual magnitude of the parameters.

Thus we have 4 possibilities for the value of the next qualitative state as follows:

$$\begin{aligned}
S^{P4P4I2I4}(t_1) &= \{ < h_l, inc > \;,\; < (T_I, T_{op}), inc > \} \\
S^{P4P4I4I2}(t_1) &= \{ < (0, h_l), inc > \;,\; < T_{op}, inc > \} \\
S^{P4P4I4I3}(t_1) &= \{ < (0, h_l), inc > \;,\; < T_{op}, std > \} \\
S^{P4P4I4I8}(t_1) &= \{ < (0, h_l), inc > \;,\; < T*, std > \}
\end{aligned}$$

Note that only the first of the above states corresponds to a normal or expected behaviour and is thus desirable. The second possible state occurs, for example, when the heating is too strong compared to the inlet flow and allows a small amount of water to boil before the amount is enough.

Having finished the first two steps of the qualitative simulation algorithm, the resulting states can be arranged in the behaviour tree shown in Fig. 7.4. It is clear that branching only occurs in the second I-transition step and the constraints cut some of the branches.

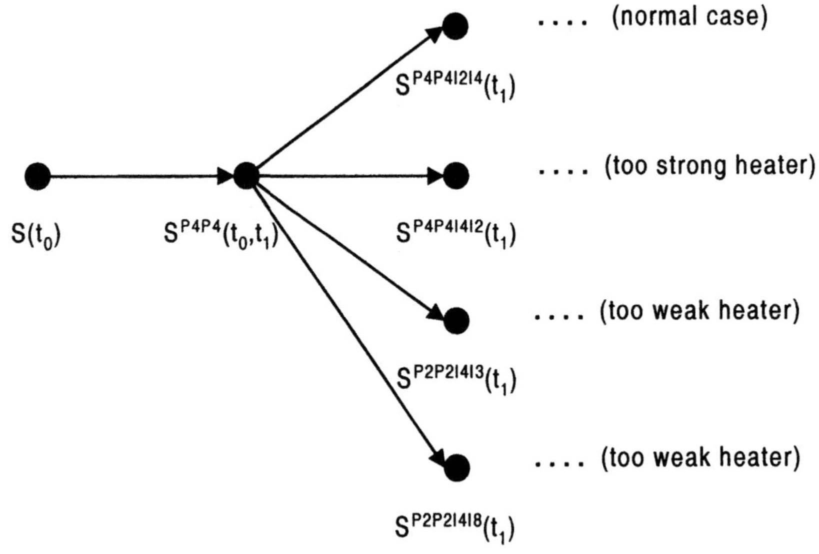


*Figure 7.4.* The behaviour tree of the batch water heater

---

Another more realistic example, model-based generation of operation procedures for a distillation column is found in [67].

# 3. QUALITATIVE PHYSICS

Qualitative physics works with sign-valued variables and qualitative differential and algebraic equations based thereon. Fist we examine these qualitative model equations and their solutions and then discuss their use in intelligent control systems.

## 3.1 CONFLUENCES

The notion of confluences as a kind of qualitative models has been introduced into AI by de Kleer and Brown [61] in their theory called "Qualitative Physics". *Confluences can be seen as sign versions of the (lumped) nonlinear state equations (7.1)-(7.2).*

They can be *formally derived* from lumped process model equations using the following steps.

1. *define qualitative variables* $[q]$ *and* $\delta q$ *to each of the model variables* $q(t)$ *as follows:*

$$q \sim [q] = sign(q) \quad , \quad dq/dt \sim \delta q = sign(dq/dt)$$

   where $sign(.)$ stands for the sign of the operand;

2. *operations are replaced by sign operations*, i.e.

$$+ \sim \oplus_S \quad , \quad * \sim \otimes_S \quad etc.$$

3. *parameters are replaced by* $+$ *or* $-$ *or* 0 forming *sign constants* in the confluence equations, i.e. they virtually disappear from the equations.

The *solution of a confluence* is computed by simply enumerating all possible values of the qualitative variables compatible with the confluence and arranging them in an *operation (truth) table of the confluence.* The operation (truth) table of a confluence contains all the possible values that satisfy it and resembles the operation table for sign operations.

It is important to note, however, that the value of the variables does change in time, therefore we have to consider various rows changing-in-time of the operation table of the confluence as time goes on.

The concepts above are illustrated on the example of the batch water heater (coffee machine) introduced in Appendix B.

**EXAMPLE 7.4** Confluences of the batch water heater of Example B

The confluences are derived from the model equations (B.1)-(B.2) in the following steps:

1. *qualitative variables*
   From their physical meaning the sign value of the variables is as follows:

$$[\eta_I] \in \{0,+\} \quad , \quad [\eta_O] \in \{0,+\} \quad , \quad [\kappa] \in \{0,+\}$$

2. *sign constants*
   The sign of all parameters is strictly positive, i.e. all sign constants are "+".

3. *confluences*

$$\delta h = [\eta_I] \ominus_S [\eta_O] \tag{7.29}$$

$$\delta T = [T_I - T] \otimes_S [\eta_I] \oplus_S [\kappa] \tag{7.30}$$

The truth table of the confluence (7.29) is shown in Table 7.6.

| $\delta h$ | $[\eta_I]$ | $[\eta_O]$ |
|---|---|---|
| 0 | 0 | 0 |
| $-$ | 0 | + |
| + | + | 0 |
| ? | + | + |

*Table 7.6.* **Truth table of a confluence on height**

The truth table of the other confluence (7.30) has more columns on the right-hand side because we have more variables there. Observe that a composite qualitative variable $[T_I - T]$ is present in the first right hand column of Table 7.7.

| $\delta T$ | $[T_I - T]$ | $[\eta_I]$ | $[\kappa]$ |
|---|---|---|---|
| + | 0 | 0 | + |
| + | + | 0 | + |
| + | − | 0 | + |
| + | 0 | + | + |
| + | + | + | + |
| ? | − | + | + |
| 0 | 0 | 0 | 0 |
| 0 | + | 0 | 0 |
| 0 | − | 0 | 0 |
| 0 | 0 | + | 0 |
| + | + | + | 0 |
| ? | − | + | 0 |

*Table 7.7.* **Truth table of a confluence on temperature**

## 3.2 THE USE OF CONFLUENCES IN INTELLIGENT CONTROL SYSTEMS

Confluences are used in intelligent control systems for two different purposes in two different ways.

1. *Sensor validation*
   If the truth table of a confluence is stored in advance then it is possible to check the value of the related sensors against the corresponding row of the table in a quick and effective way. A sensor fault for the group of related sensors is detected if there is a contradiction between the right-hand side of the corresponding row in the table and the measured qualitative value of the left-hand side sensor.

2. *Knowledge base items*
   The rows of the truth table of a confluence can be interpreted as a rule if one reads them from right to left.

   For example, the second row in table 7.6 is interpreted as

$$(\eta_I = 0) \wedge (\eta_O = 1) \quad \rightarrow \quad (\delta h = -)$$

or

$$\textbf{if} \ \ (\eta_I = \textbf{closed}) \ \textbf{and} \ (\eta_O = \textbf{open}) \ \ \textbf{then} \ \ (h = \textbf{decreasing}) \tag{7.31}$$

This way *rule sets can be generated from the truth table of a confluence and they are complete and contradiction-free by construction.*

Note that in some cases **if-and-only-if** type bidirectional rules or rule-pairs can be generated from a row in a truth table. This is possible if the left-hand side of that row is unique among the left-hand side values.

For example, let us assume that the last row is missing from table 7.6. The omission is needed to make the values of the left-hand side unique, leaving out "?", which is the union of "+", "0" and "−". Then each row in this new 3-row generates a bidirectional rule-pair. The second row is interpreted as a rule (7.31) and its counterpart in the form:

$$\textbf{if}\ \ (h = \textbf{decreasing})\ \ \textbf{then}\ \ (\eta_I = \textbf{closed})\ \textbf{and}\ (\eta_O = \textbf{open}) \tag{7.32}$$

# 4. SIGNED DIRECTED GRAPH (SDG) MODELS

Signed directed graph (SDG) models are the most rude qualitative models: they basically describe the structure of the linearized state-space model of a dynamic system.

## 4.1 THE STRUCTURE GRAPH OF STATE-SPACE MODELS

The structure of a continuous-time deterministic dynamic system given in linear time-invariant or time-varying parameter state-space form with the equations (7.1)-(7.2) or (7.3)-(7.4) can be represented by a *directed graph* (see Murota, 1987 [68]; Reinschke, 1988 [62]) as follows. The nodes of the directed graph correspond to the system variables; a directed edge is drawn from the $i$th vertex to the $j$th one if the corresponding matrix element is not equal to zero. Hence, if the $i$th variable is present on the right-hand side of the $j$th equation then an edge exists.

The *system structure* above is represented by a *directed graph* $S = (V, \mathcal{E})$ where the vertex set $V$ is partitioned into three disjoint parts,

$$\begin{gathered} V = X \cup U \cup Y \\ X \cap U = X \cap Y = U \cap Y = \emptyset \end{gathered} \tag{7.33}$$

with $X$ being the set of $q$ state variables, $U$ the set of $p$ input variables and $Y$ the set of $z$ output variables. All edges $e \in \mathcal{E}$ terminate either in $X$ or in $Y$, by assumption:

$$v_2 \in (X \cup Y) \quad \forall\, e = (v_1, v_2) \in \mathcal{E} \ ,$$

i.e. there are no inward directed edges to input variables. Moreover, all edges start in $(U \cup X)$, again by assumption:

$$v_1 \in (X \cup U) \quad \forall\, e = (v_1, v_2) \in \mathcal{E} \ ,$$

That is, there are no outward directed edges from outputs of the graph. The graph $S$ is termed the ***structure graph of the dynamic system***, or ***system structure graph.***

Sometimes, the set $U$ is called the *entrance* and $Y$ the *exit* of graph $S$.

***Directed paths*** in a system structure graph can be used to describe the effect of a variable on other variables. A sequence $(v_1, \cdots, v_n)$ , $v_i \in V$, forming a directed path in $S$, is an ***input path*** if $v_1 \in U$.

We may want to introduce a more general model, taking into account that the connections between the elements of a process system structure are not necessarily of the same strength. To represent such conditions we consider *weighted digraphs*, which give full information not only on the structure matrices $[W]$ but also on the entries of $W$. In this way the actual values of the elements in state-space representation matrices $(A,\ B,\ C,\ D)$ can be taken into account.

To do this, we define a weight function $w$ whose domain is the edge set $\mathcal{E}$ of the structure graph $S = (V, \mathcal{E})$, and assign the corresponding matrix element in $W$ as weight

$$w(v_i, v_j) := w_{ij} \tag{7.34}$$

to each edge $(v_i, v_j)$ of $S$.

For *deterministic state-space models without uncertainty*, the edge weights $w_{ij}$ are real numbers. In this way, a one-to-one correspondence is established between the state-space model (7.3)-(7.4) and the weighted digraph $S = (V, \mathcal{E}; w)$.

We may may put the sign of the corresponding matrix elements ($\{A\}$, $\{B\}$, $\{C\}$, $\{D\}$) as the weights $w_{ij}$. We arrive at a special weighted directed graph representation of the process structure, which is called the ***signed directed graph (SDG)*** model of the process system.

**EXAMPLE 7.5** SDG model of the batch water heater introduced in Appendix B

The state-space model of the batch water heater consists of the state equations (B.1)-(B.2) and the trivial output equation

$$y = x = [h \ T]^T$$

We can then linearize the state equations taking into account that our input variables are $u = [\kappa, \ \eta_I, \ \eta_O]^T$.

In order to distinguish between the partitions of the SDG model vertices, circles will be applied for the state, double circles for the input and rectangles for the output variables. With this notation the SDG model of the batch water heater (coffee machine) is shown in Fig. 7.5.

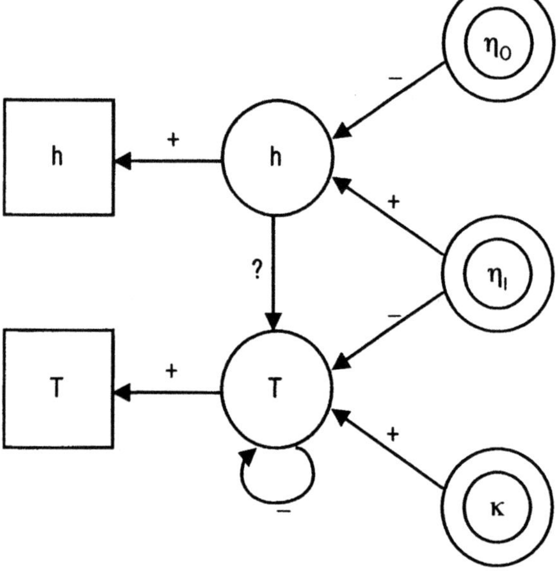


*Figure 7.5.* SDG of the coffee machine

## 4.2 THE USE OF SDG MODELS IN INTELLIGENT CONTROL SYSTEMS

Structure graphs and SDG models are simple and transparent tools that represent the structure of dynamic models. They not only represent a particular system they have been derived for but a whole *class of process systems* with the same structure. Moreover, they are highly modular and able to capture the system-subsystem hierarchy in a transparent way. These properties and their simplicity makes these tools very useful in intelligent control applications.

There are two basic application areas for SDG models.

1. *Analysing dynamic properties of a class of systems*
   This is not the main application area relevant to intelligent control, therefore we only give a list of the most important uses of structure graphs and SDGs:

   - *Analysis of structural controllability and observability* [62] by investigating the input and output connectivity of a structure graph.
   - *Analysis of structural stability* [69] by computing the sign values of circles and circle families in a graph.
   - *Qualitative analysis of unit step responses* by evaluating the sign value of the shortest path(s) and that of the circles and circle families [70].

2. *Diagnostic reasoning* [71], [72]
   An SDG can be seen as a special type of influence graphs that describes the cause-consequence relationship between the deviation of signals from their nominal value. Assume, for example, that we have an edge $A \rightarrow B$ in our SDG between two system variables $A$ and $B$ and the sign associated with that edge is denoted by $sgn(A - B)$. Then the sign of the deviation of the variable $B$ (more precisely the deviation of the signal associated with it from its nominal value) can be computed by

   $$sgn(B) = sgn(A - B) \cdot sgn(A)$$

   where $sgn(A)$ is the deviation of the variable $A$. If we have a path instead of just an edge connecting the two variables such as $P(A, B) = (A, x_1, \ldots, x_n, B)$ then

   $$sgn(B) = sgn(x_n - B) \cdot \cdots \cdot sgn(A - x_1) \cdot sgn(A)$$

In this case, the sign of the deviation of a target variable can be computed by forward reasoning along directed paths starting from the vertices where measured variables are found.

The diagnosis based on the qualitative model uses the SDG models for various faulty modes or situations and compares the qualitative prediction performed by these models with the measured reality. Fault detection and isolation is then performed by simple comparison.

Chapter 8

# PETRI NETS

Petri nets are the abstract formal models of information streams. They were named after C. A. Petri, a German mathematician, who developed the method for modelling communication of automata [73].

Petri nets allow both the mathematical and the graph representation of a discrete event system to be modelled, where the signals of the system have discrete range space and time is also discrete. Petri nets can be used for describing a controlled or open-loop system, for modelling the events occurring in it and for analyzing the resulted model. During the modelling and analysis process we can get information about the structure and dynamic behaviour of the modelled system.

Petri nets emphasize the possible states and the events occurring in the investigated system. One of the major strengths of Petri nets is in the modelling of parallel events. This is why Petri nets are popular and widely used discrete dynamic modelling tools in intelligent control applications [74], [75], [76], [77].

This chapter discusses the following topics:

- the notion of Petri nets,
- the dynamical behaviour of Petri nets,
- the analysis of Petri net properties.

We mainly focus on the modelling problems and questions related to the analysis of Petri nets. The analysis techniques need software support even with low complexity systems.

Throughout this chapter the following simple process system is used for the demonstration of the modeling capabilities of Petri nets.

---

**EXAMPLE 8.1** A simple process system

Let us look at a simple jacketed reactor which has heating - cooling capabilities and which is equipped with a stirrer. In the reactor, a simple endotherm reaction takes place in which reagents $A$ and $B$ react with each other in a given ratio forming a solid product $C$. During the reaction the temperature of the reactor has to be kept at a given value and the stirrer has to operate continuously. The raw materials are stored in smaller tanks for daily use and they are pumped from these tanks to the reactor. At the end of the reaction the content of the reactor has to be cooled down, the product has to be filtered and the solvent can be recirculated.

---

# 1. THE NOTION OF PETRI NETS

## 1.1 THE BASIC COMPONENTS OF PETRI NETS

In the following we introduce the basic elements of Petri nets first using some introductory examples, then the formal definition is given [78], [79].

### 1.1.1 INTRODUCTORY EXAMPLES

A Petri net consists of ***places*** and ***transitions***, and describes the relations between them. As the names of these elements show, places refer to static parts of the modelled system while transitions refer to changes or events occurring in the system.

The mathematical representation of Petri nets consists of the sets of transitions and places, the functions describing the relations between them and a function that describes the dynamic state of the net. The graphic representation of Petri nets is a bipartite directed graph where places are drawn as circles and transitions are drawn as bars or boxes. Logical relations between transitions and places, i.e. between events and their preconditions and consequences are represented by directed arcs.

Transitions can be seen as the steps or substeps of operating procedures while places imply the preconditions and consequences of these steps in a controlled discrete event system.

In a complex system a consequence of an event is a precondition of other events. We generally use the term 'condition' instead of both precondition and consequence. In a Petri net model we use the expressions *input place* and *output place* rather than precondition and consequence if it is necessary to emphasize the relation between a place and a transition.

The validity (or occurrence) of a condition in the modelled system can be represented by the presence or absence of tokens in the appropriate place in the net or by nonnegative numbers (in the mathematical representation). If the condition is not valid in the real system, then there is no token in its place or the value associated to it is equivalent to zero. On the other hand, if the condition is valid, then there is a token in its place or its value is equivalent to one. In certain cases, there can be more than one token in a given place.

In the so-called low-level Petri nets there is no distinction made between the tokens. This means that items represented by tokens in a given place are either the same or the differences between them are not relevant from the point of view of the modelling goal.

---

**EXAMPLE 8.2** Basic elements of Petri nets

Consider the reactor of the simple process system described in Example 8.1 where we want to pump out its content. Place $p_{ready}$ represents the state of the reactor. If there is a token in this place then the reaction is over and the reactor is ready to be emptied. If the reactor does not contain material to be emptied then this place does not contain token.

Place $p_{pump}$ represents the availability of a pump and a token means a pump is available. If there is no token in that place then there is no free pump in the system. Transition $t_{drain}$ refers to the pumping out operation while place $p_{tank}$ belongs to the state of the tank (a container) containing the product. The Petri net of this system can be seen in Fig. 8.1.

Assume we have two pumps to perform this operational step and there is no constraint which one we use. The number of tokens in place $p_{pump}$ represents the number of available pumps. Two or more tokens represent the availability of two or more pumps at the same time as it is seen in Fig. 8.2. If we want to distinguish the pumps we have to assign a separate place to each pump and define separate operating steps as depicted in Fig. 8.3.

---

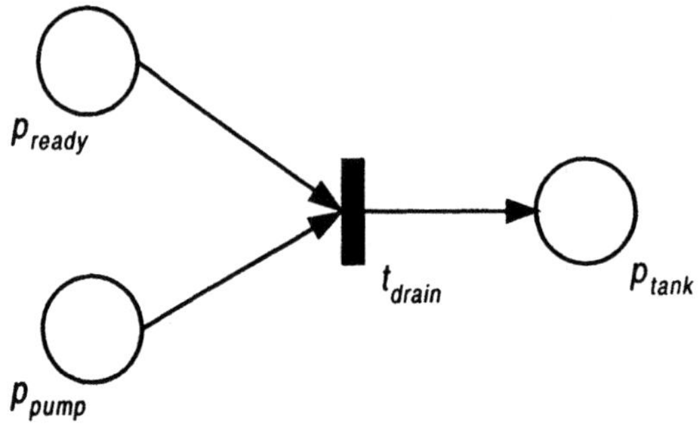


*Figure 8.1.* Drain section of the process example

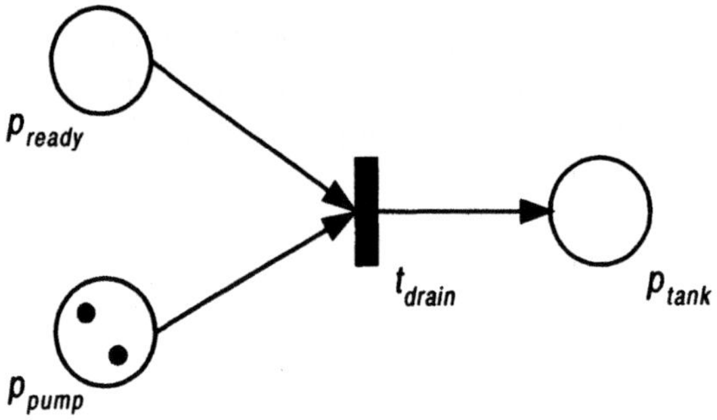


*Figure 8.2.* A drain section with two pumps

The fact that a place contains more than one token can have quite different meanings. In the following example, we use tokens as a quantity indicators (measures).

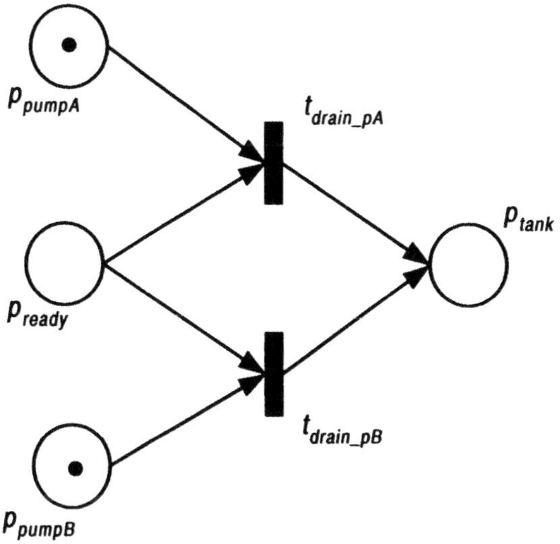


*Figure 8.3.* A drain section with two distinguished pumps

---

**EXAMPLE 8.3** Places with more than one tokens

Let us model the feeding part of the process system described in Example 8.1 the introduction of this chapter. We have to feed the reactor with a mixture containing reagents $A$ and $B$ in ratio 1:1. The graph of this net can be seen on Fig. 8.4. Place $p_{rA}$ refers to the back storage tank of component $A$ and place $p_{rB}$ to the back storage tank of component $B$. Assume we have a large amount of material in both tanks, so there are several tokens in both places referring to the tanks. Places $p_A$ and $p_B$ represent daily storage tanks for adding the reagents to the reactor.

If we assume that one token refers to the amount of reagents $A$ and $B$ to be added into the reactor for one charge then the maximum number of tokens on these places is equal to 1.

The place $p_{filled}$ refers to the state of the reactor. It contains a token when both reagent $A$ and reagent $B$ are filled. Transitions $t_{addA}$ and $t_{addB}$ refer to the filling up of the daily tanks and transition $t_{fill}$ represents the reactor filling up process.

---

The modelling of the investigated system using places, transitions and arcs means the description of the static structure of the system. This is

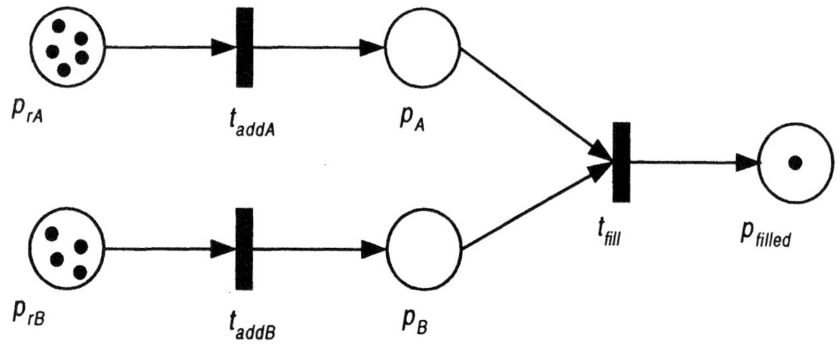


*Figure 8.4.* The feeding part of the process example

useful in itself because it helps the understanding of the system structure and it is also used in the analysis. At the same time, the Petri net we got as a result makes it possible to carry out behavioural investigations, when we want to know what will happen in the system starting from an initial state.

The firing of transitions in the net follows the behaviour of the real system: an event can occur if all of its preconditions are fulfilled. In the Petri net, a transition is called enabled if if all of its input places are valid. If a transition is enabled then it can fire. If an event occurs in the real system then the transition referring to this event must be fired in the net.

During the firing of a transition the appropriate number of tokens is removed from the input places and added to the output places. The logical relations between places and transitions define the number of tokens to be removed or added. This process is illustrated by the following example.

---

**EXAMPLE 8.4** The firing of a transition

Here we model the reaction part of our system described in Example 8.1. The Petri net of this part can be seen in Fig. 8.5.

In this figure place $p_{filled}$ refers to the fact that the reactor is ready for heating up and the reaction, place $p_{react}$ refers to the reaction and $p_{ready}$ is associated with the state after the end of the reaction. The transitions $t_{heat}$ and $t_{cool}$ refer to heating up and cooling down.

Assume the content of the reactor is ready to be heated up. This is the initial state, i.e. there is a token in the place $p_{filled}$.

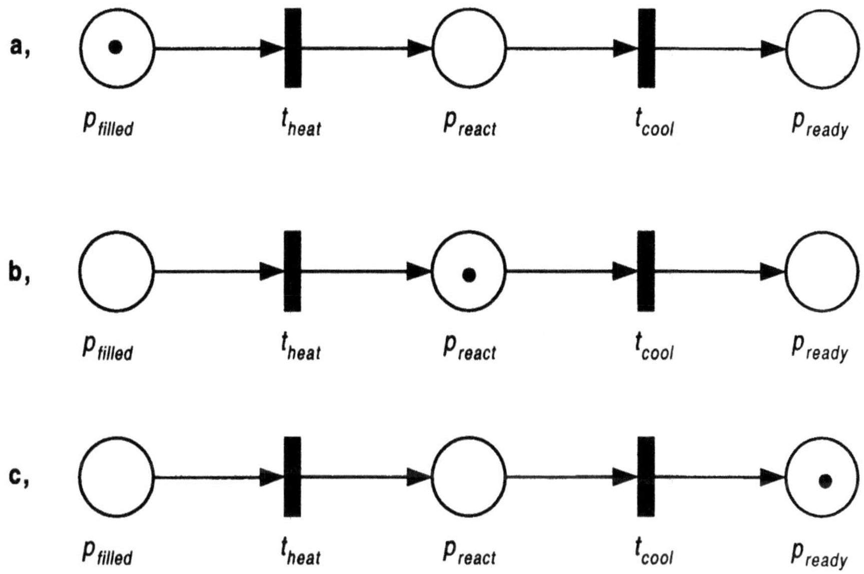


*Figure 8.5.* **The reaction part of the process example**

There is only one transition in the initial state, transition $t_{heat}$, which is enabled (c.f. Fig. 8.5.a). Firing transition $t_{heat}$ removes the token from its input place (place $p_{filled}$) and adds this token to its output place (place $p_{react}$) (c.f. Fig. 8.5.b). In the next step, transition $t_{cool}$ is the only enabled transition. Following the rule for firing transitions we get to the state in Fig. 8.5.c.

In the resulting state there is no enabled transition in the net so the execution of the net starting from the given initial state is over.

---

As we mentioned earlier the number of added and removed tokens in a place depends on the nature of logical relations between the given place and its neighbouring transitions. In the following example we show when and how can we handle this case.

**EXAMPLE 8.5** Arcs with weights

Let us once more investigate the adding part in Example 8.3. Let us assume that we have to add one portion of reagent $A$ and two portions of reagent $B$ to the mixture. The reactor is ready for heating up if the feeding is done, i.e. all the necessary components have been added. The modified Petri net of the feeding part can be seen in Fig. 8.6.

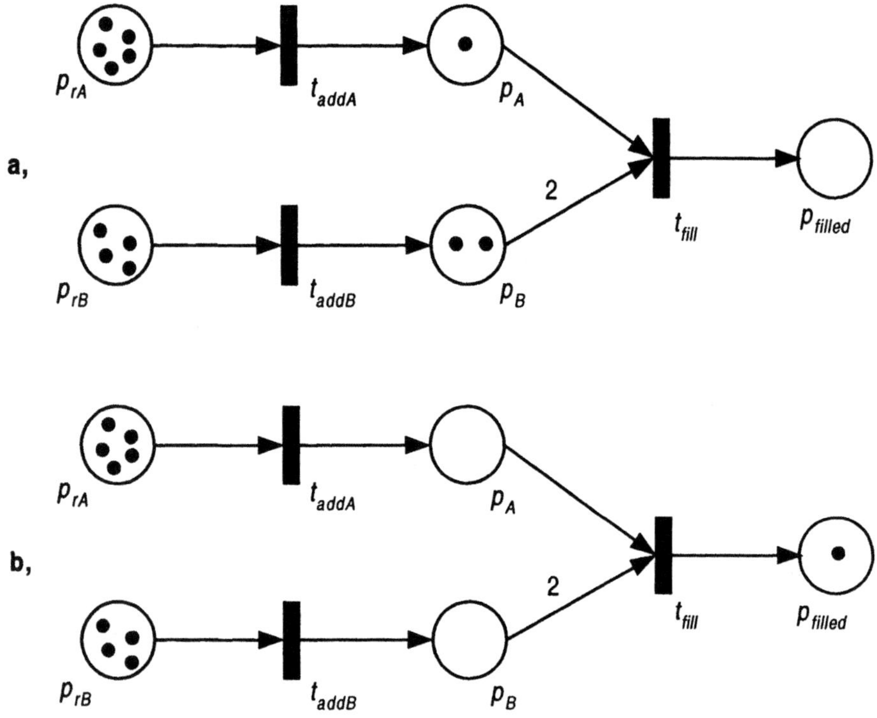


*Figure 8.6.* The modified feeding part of the process example

In this figure the arc between place $p_A$ and transition $t_{fill}$ is the same as earlier, while the arc between place $p_B$ and transition $t_{fill}$ is coloured by a weight with a value of 2. This weight means that transition $t_{fill}$ is enabled if its input place $p_B$ contains at least two tokens. Firing transition $t_{fill}$ removes one token from place $p_A$, two from place $p_B$ and adds one to place $p_{filled}$ as it is shown in Fig. 8.6.b.

In general, a $k$-weighted arc can be interpreted as $k$ parallel arcs between the given place and transition pair. This example shows that the tokens don't travel from place to place in the net but the actual number of tokens in a place is calculated on the basis of the logical relations defined between places and transitions.

As we have seen the number of tokens may change during the execution of the net. Obviously, the distribution of tokens in the places characterizes the state of the net. We can use a marking function to assign the appropriate token value to places. The marking can be interpreted as a vector which has $m$ components where $m$ is equal to the number of places. In case of a given place the $M(p_i)$ form is used to give the actual number of tokens in place $p_i$. In general, the marking $M_0$ refers to the initial state.

---

**EXAMPLE 8.6** Markings

The connection of the feeding and reaction parts of our process system in Example 8.1 results in the Petri net shown in Fig 8.7.

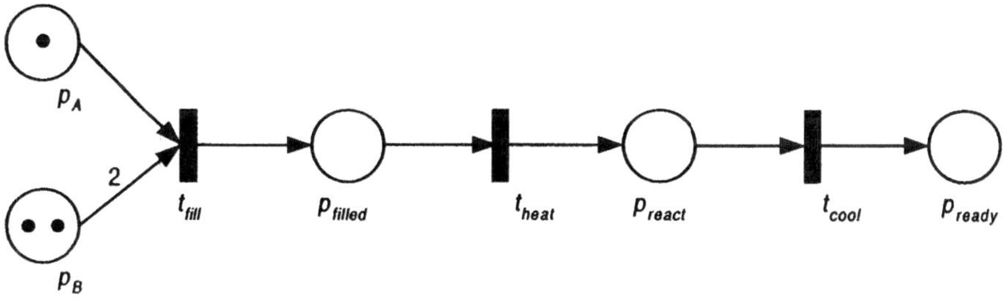


*Figure 8.7.* The feeding and reaction parts of the process example

We choose the initial marking to be

$$M_0(p_A, p_B, p_{filled}, p_{react}, p_{ready}) = (1, 2, 0, 0, 0) . \tag{8.1}$$

in this example.

After firing transition $t_{fill}$ the new marking is

$$M_1(p_A, p_B, p_{filled}, p_{react}, p_{ready}) = (0, 0, 1, 0, 0) . \tag{8.2}$$

As the next step we start the heating process (see Example 8.4) then the result is the following marking

$$M_1(p_A, p_B, p_{filled}, p_{react}, p_{ready}) = (0, 0, 0, 1, 0) . \quad (8.3)$$

---

### 1.1.2 THE FORMAL DEFINITION OF PETRI NETS

Based on the introductory examples of the previous part in this section the formal definition of Petri nets is the following:

**Definition 8.1.** *A Petri net is a 4-tuple*

$$N = < P, T, F, W >, \quad (8.4)$$

*where*

$P \ = \{p_1, p_2, \dots, p_m\}$ *is a finite set of places;*
$T \ = \{t_1, t_2, \dots, t_n\}$ *is a finite set of transitions;*
$F \subseteq (P \times T) \cup (T \times P)$ *is a set of arcs;*
$W : F \rightarrow \mathbb{N}$ *is a weight function*
*where* $\mathbb{N}$ *is a set of nonnegative integers;*
*furthermore* $P \cap T = \emptyset$ *and* $P \cup T \neq \emptyset$.

*The marking function gives the distribution of tokens in a given net state:*
$M : P \rightarrow \mathbb{N}$

*A Petri net with a given initial marking is denoted by*
$PN = < N, M_0 >$ ■

Note that a Petri net is said to be *ordinary* if all of its arc weights are equal to 1.

## 1.2 THE FIRING OF TRANSITIONS

In the previous section we have introduced the basics of firing transitions using some simple examples. In the following this question is discussed in detail.

The **firing rules of transitions** are the following:

1. A transition $t_j$ is said to be enabled if there is at least $w(p_i, t_j)$ token on each input place $p_i$ of $t_j$:

$$M(p_i) \geq w(p_i, t_j) \qquad \text{for } \forall p_i \in P \quad (8.5)$$

where $w(p_i, t_j)$ is the arc weight.

2. An enabled transition may or may not fire depending on whether or not the event modelled by the transition actually takes place in the real system.

3. At firing of transition $t_j$ the value of the marking function of a place is decreased by the weight of the arc connecting the given place to transition $t_j$, and is increased by the weight of the arc from transition $t_j$ to the given place:

$$M''(p_i) = M'(p_i) - w(p_i, t_j) + w(t_j, p_i) \qquad \text{for } \forall p_i \in P \qquad (8.6)$$

As it can be seen we generalized the increases and decreases for all places in the net. These operations have no effect on the marking value if there is no logical relation between the given place and transition so this generalization enables a simpler treatment of markings.)

As an example for the practical application of these rules we repeat the first two steps of Example 8.6.

**EXAMPLE 8.7** The calculation of marking values

The initial marking in Example 8.5 is equal to

$$M_0(p_A, p_B, p_{filled}, p_{react}, p_{ready}) = (1,2,0,0,0)\,. \tag{8.7}$$

In this initial state the transition $t_{fill}$ is the only enabled transition. After its firing the new marking can be given by the following equations:

$$\begin{aligned} M_1(p_A) &= M_0(p_A) - w(p_A, t_{fill}) + w(t_{fill}, p_A) \\ M_1(p_A) &= 1 - 1 + 0 = 0 \end{aligned} \tag{8.8}$$

$$\begin{aligned} M_1(p_B) &= M_0(p_B) - w(p_B, t_{fill}) + w(t_{fill}, p_B) \\ M_1(p_B) &= 2 - 2 + 0 = 0 \end{aligned} \tag{8.9}$$

$$\begin{aligned} M_1(p_{filled}) &= M_0(p_{filled}) - w(p_{filled}, t_{fill}) + w(t_{fill}, p_{filled}) \\ M_1(p_{filled}) &= 0 - 0 + 1 = 1 \end{aligned} \tag{8.10}$$

$$\begin{aligned} M_1(p_{react}) &= M_0(p_{react}) - w(p_{react}, t_{fill}) + w(t_{fill}, p_{react}) \\ M_1(p_{react}) &= 0 - 0 + 0 = 0 \end{aligned} \tag{8.11}$$

$$\begin{aligned} M_1(p_{ready}) &= M_0(p_{ready}) - w(p_{ready}, t_{fill}) + w(t_{fill}, p_{ready}) \\ M_1(p_{ready}) &= 0 - 0 + 0 = 0 \end{aligned} \tag{8.12}$$

Summarizing the equations the firing of transition $t_{filling}$ results in marking $M_1 = (0,0,1,0,0)$. In case of places $p_{react}$ and $p_{ready}$ the zero values of the weight function $w$ express that the are neither input nor output connections between transition $t_{fill}$ and these places.

In marking $M_1$ only transition $t_{heat}$ is enabled and its firing results (given as marking $M_2$) are as follows.

$$\begin{aligned} M_2(p_A) &= M_1(p_A) - w(p_A, t_{heat}) + w(t_{heat}, p_A) \\ M_2(p_A) &= 0 - 0 + 0 = 0 \end{aligned} \tag{8.13}$$

$$\begin{aligned} M_2(p_B) &= M_1(p_B) - w(p_B, t_{heat}) + w(t_{heat}, p_B) \\ M_2(p_B) &= 0 - 0 + 0 = 0 \end{aligned} \tag{8.14}$$

$$\begin{aligned} M_2(p_{filled}) &= M_1(p_{filled}) - w(p_{filled}, t_{heat}) + w(t_{heat}, p_{filled}) \\ M_2(p_{filled}) &= 1 - 1 + 0 = 0 \end{aligned} \tag{8.15}$$

$$\begin{aligned} M_2(p_{react}) &= M_1(p_{react}) - w(p_{react}, t_{heat}) + w(t_{heat}, p_{react}) \\ M_2(p_{react}) &= 0 - 0 + 1 = 1 \end{aligned} \tag{8.16}$$

$$\begin{aligned} M_2(p_{ready}) &= M_1(p_{ready}) - w(p_{ready}, t_{heat}) + w(t_{heat}, p_{ready}) \\ M_2(p_{ready}) &= 0 - 0 + 0 = 0 \end{aligned} \tag{8.17}$$

---

## 1.3 SPECIAL CASES AND EXTENSIONS

In this section some special cases of firing transitions and some extensions to the original Petri net structure will be introduced.

### 1.3.1 SOURCE AND SINK TRANSITIONS

A transition without any input place is called a *source transition.* A source transition can fire in any net state, i.e. it is unconditionally enabled.

A transition without any output place is called a *sink transition.* If a sink transition fires, the amount of tokens decreases because this type of transition consumes tokens without producing them.

### 1.3.2 SELF-LOOP

If a place is an input *and* output place of the same transition then this place-transition pair is called a *self-loop.* If the weights are the same in both cases then the firing of the transition does not change the marking value on that place. It can be proved that if the place belonging to the loop cannot be found in the input set of any other transition and the transition has no other input place then once the transition becomes enabled it remains enabled throughout the execution of the net. A Petri net without self-loop is said to be *pure*.

The following example gives a process system illustration.

---

**EXAMPLE 8.8** Source and sink transitions and a self-loop

A source transition represents an event or operation which occurs in every step. It can be the continuous arrival of new parts on a conveyor belt as it can be seen in Fig. 8.8.

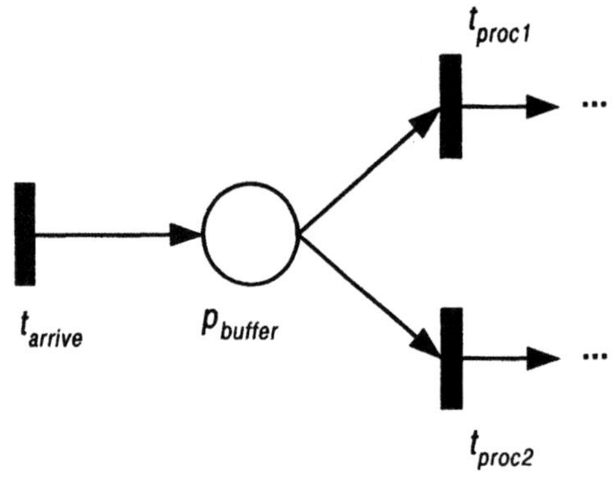


*Figure 8.8.* Modelling of continuous arrival

A sink transition can represent the shipping of the product into the depository or the intermediates into another unit. Its Petri net model can be seen in Fig. 8.9.

In general, these transitions are used when modelling a subsystem of complex process systems.

The self-loop can refer to a continuous operation of a device, e.g. of a stirrer, which is a precondition of every process step (c.f. Fig. 8.10).

---

### 1.3.3 CAPACITY OF PLACES

Up to this point we assumed that places can have an infinite number of tokens, i.e. there is no constraint on the marking value of a place.

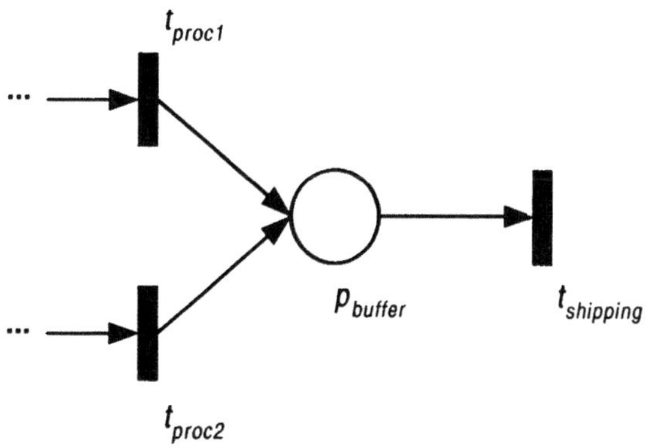


*Figure 8.9.* The modelling of shipping

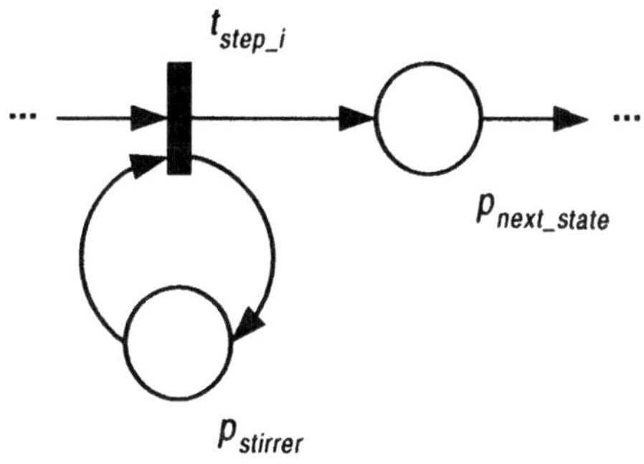


*Figure 8.10.* Modelling a constant precondition

A Petri net of this type is called an *infinite capacity net.* However, real discrete event systems are different.

In Example 8.3 the reactor cannot contain more than one token, because the presence of one token means that the reactor is full. This means we have to add an upper limit for the number of tokens for every place. Such a Petri net is referred to as a *finite capacity net.* For a finite capacity net we have to add a new function, $K$, which is the capacity function to the formal definition:

$$K : P \to \mathbb{N}^+ \ , \tag{8.18}$$

and we have to modify the rule for transitions to be enabled.

A transition $t_j$ is said to be enabled if there is at least $w(p_i, t_j)$ token in each input place $p_i$ of $t_j$:

$$M'(p_i) \geq w(p_i, t_j) \qquad \text{for } \forall p_i \in P \tag{8.19}$$

and after the firing of $t_j$ the marking in its output places does not exceed their upper limits:

$$M''(p_i) \leq K(p_i) \qquad \text{for } \forall p_i \in P$$
$$\text{where } M''(p_i) = M'(p_i) - w(p_i, t_j) + w(t_j, p_i). \qquad (8.20)$$

This modified rule is called a ***strict transition rule*** whereas the former rule for an infinite capacity net is the ***(weak) transition rule.***

### 1.3.4 PARALLELISM

One of the most important modelling feature of Petri nets is the handling of parallelism. If the steps of an operating procedure take place in a sequence of elementary steps, i.e. the system has a serial nature then the model is easily understandable. But real systems almost always contain serial ***and*** parallel steps, this is even true for simpler discrete event systems. In the case of parallel steps one of the most important task is timing.

There are two different types of parallelism: *concurrency* and *conflict.* In the case of concurrency, two (or more) events take place in parallel. They can occur at the same time because they are causally independent. This means that one transition may fire before, after or in parallel with the other as it can be seen in the following example.

---

**EXAMPLE 8.9** Concurrency

Let us amplify Example 8.3 with the start of the stirrer (see Fig. 8.11). As it can be seen transitions $t_{fill}$ and $t_{stir\_start}$ can fire independently of each other in any order.

---

In the other case, the events in parallel are not causally independent, i.e. they have at least one common precondition. This means that only one of these transitions can fire because after the firing of the first one, the other transition(s) is/are not enabled. The events or the transitions referring to them in the net are *in conflict.*

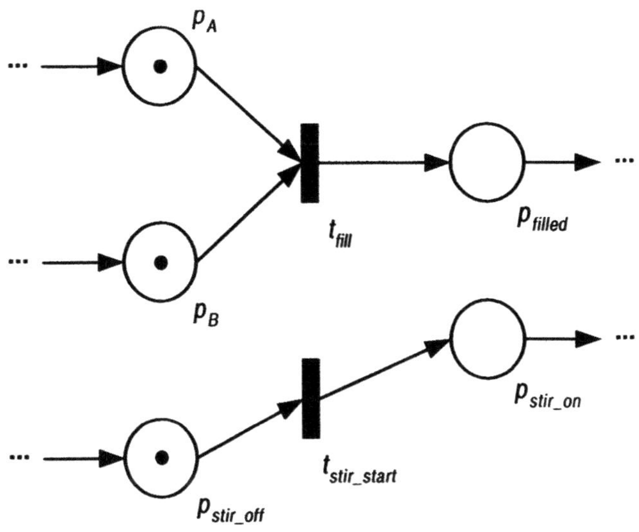


*Figure 8.11.* **Modelling concurrent events**

---

**Example 8.10** Conflict

As a next step let us modify the feeding process of the reactor in Example 8.3 in the following way (see Fig. 8.12).

We use a pump to feed the reagents to the reactor and there is only one pump which serves both reagents. Place $p_{pump}$ refers to the state of the pump. If there is a token in that place then the pump is idle otherwise it is busy. Transitions $t_{fill\_A}$ and $t_{fill\_B}$ refer to the feeding of components $A$ and $B$.

As it is seen in Fig. 8.12 either transition $t_{fill\_A}$ or transition $t_{fill\_B}$ can fire first but they cannot fire at the same time.

Assume that after the completion of the reaction, the content of the reactor must be filtered and we have two filters in the system to do this. If we can let the content of the reactor into either of them and both filters are available then both transitions ($t_{filt\_A}$ and $t_{filt\_B}$) are enabled at the same time and we have to choose between them. If filtering starts in filter $A$, i.e. transition $t_{filt\_A}$ fires then the other transition $t_{filt\_B}$ is not enabled as it is shown in Fig. 8.13.

In this case only one transition can fire.

---

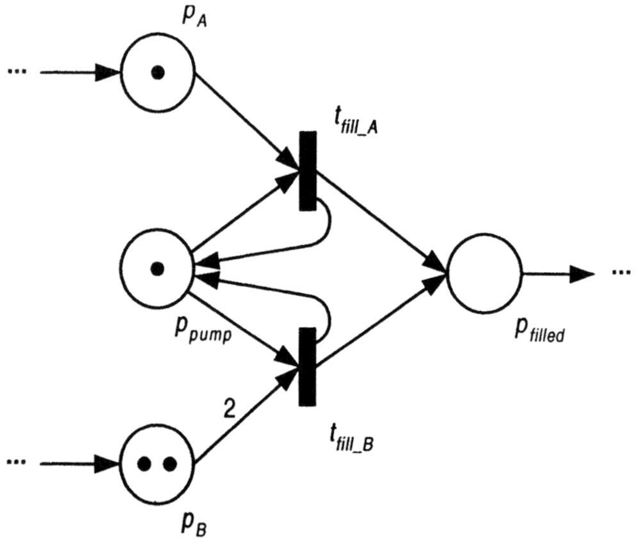


*Figure 8.12.* Modelling events in a conflict situation

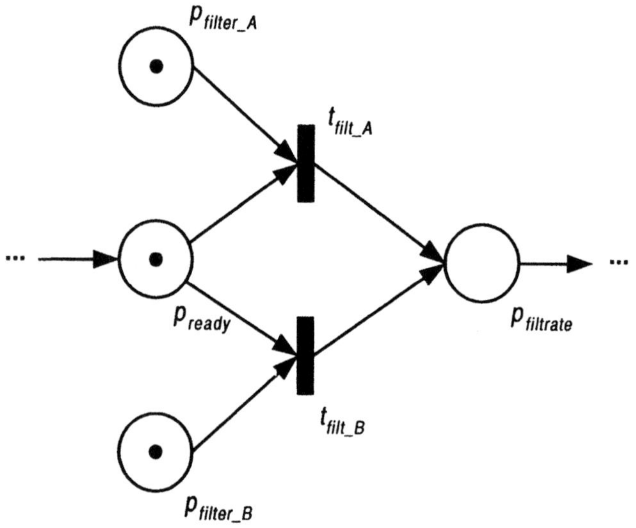


*Figure 8.13.* Conflict at filtering

The situation where conflict and concurrency are mixed is called *confusion*. There are two types of confusion: *symmetric* and *asymmetric confusion*.

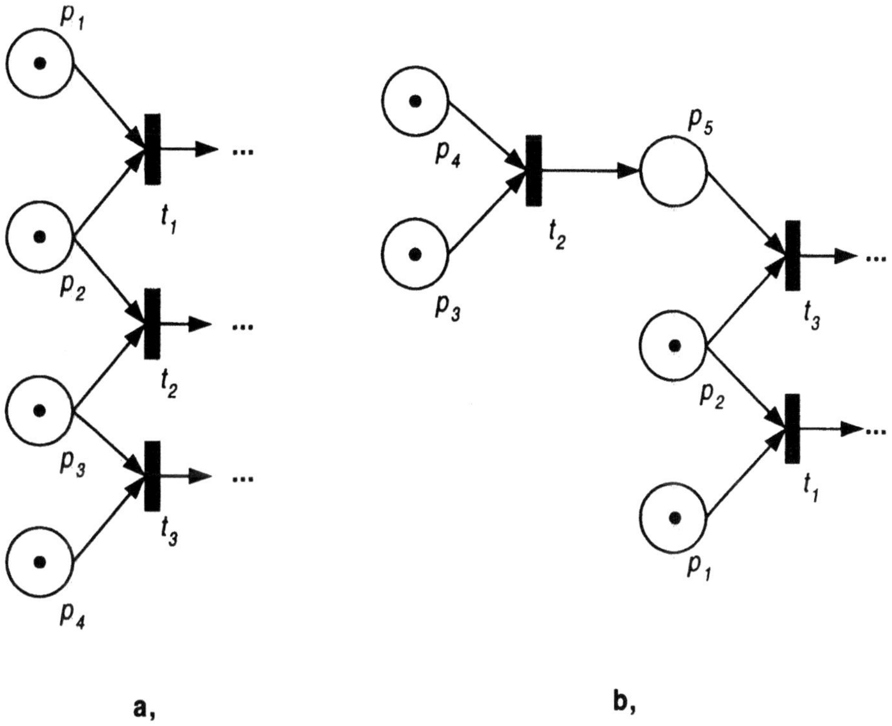


*Figure 8.14.* Symmetric and asymmetric confusion

In the case of symmetric confusion (see Fig. 8.14.a) transition $t_1$ and $t_3$ are in concurrency, both transitions can fire independently of each other. On the other hand, both transitions are in conflict with transition $t_2$ because once that fires neither of them remain enabled.

An example for asymmetric confusion can be seen in Fig. 8.14.b. In this case transitions $t_1$ and $t_2$ are in concurrency but if $t_2$ fires first the model gets into a conflict situation between transitions $t_1$ and $t_3$.

The exploration of parallel events in a system is one of the most important tasks during modelling. In the case of concurrency it can be proved that the events in a parallel situation can occur independently of each other. Synchronization of the two events has to be organized separately if that is necessary.

The presence of a conflict situation in the model can refer to the presence of uncertainty in the system. In some situations it causes no difference which transition is chosen, as we have seen in Example 8.2 in Fig.8.3 (Drain section with two distinguished pumps). But in other cases, an unfortunate selection can cause dangerous situations. Clarifying these situations helps to make the operation of the system or the operational procedure unambiguous.

### 1.3.5 INHIBITOR ARCS

As we have seen before, the firing of a transition depends on the presence of the appropriate number of tokens in its input places. But in some cases the lack of tokens in a place can be the precondition of the firing of a transition.

---

**EXAMPLE 8.11** The zero-test

Let us investigate the drain section of the process system in Example 8.1 (see Fig. 8.15.).

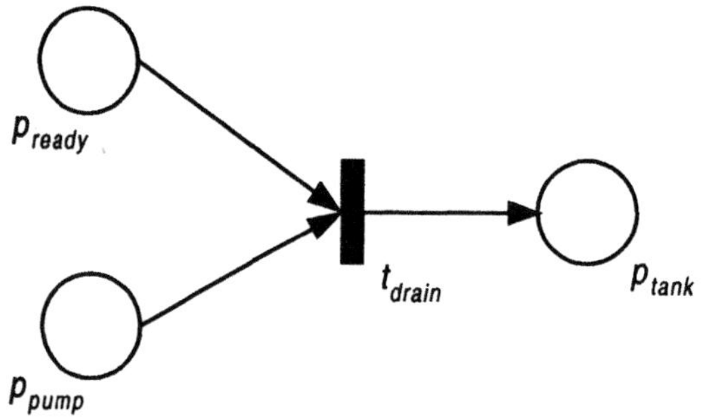


*Figure 8.15.* Draining section

The transition $t_{drain}$ is enabled if there is a token both in places $p_{ready}$ and $p_{pump}$. But we have to assume at the same time that the tank storing the product is empty, i.e. there is no token in place $p_{tank}$.

A simple solution for this problem can be seen in Fig. 8.16. Here the state of the tank is divided into two states: place $p_{tank\_full}$ refers to the state when there is material in the tank, and place $p_{tank\_empty}$ refers to the empty state. These two places are in a special relationship because one and only one of them can have the token at a time. This situation is called mutual exclusion.

---

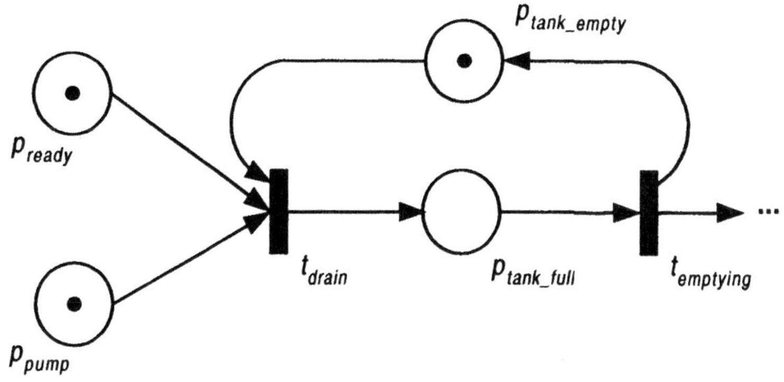


*Figure 8.16.* **Modelling of the state of tank with two places**

In these cases zero-testing is needed. For this purpose the so-called *inhibitor arc* is introduced into the Petri net modelling as an extension. An inhibitor arc is always directed from a place to a transition and has a small circle rather than an arrowhead at its endpoint.

If we allow the presence of inhibitor arcs in the model the firing rule of transitions has to be changed as follows. A transition is enabled if the number of tokens in its input places connected by arrowhead arcs is equal to or larger than the value of weight functions assigned to the arcs and there is no token in its input places connected by inhibitor arcs. The removal and addition of tokens from the input places and to output places does not change.

---

EXAMPLE 8.12 Inhibitor arcs

Directing an inhibitor arc from place $p_{tank}$ to transition $t_{drain}$ is one of the possible solutions to the problem mentioned in Example 8.11.

According to this modification transition $t_{drain}$ is enabled if and only if:

- cooling down has finished - there is a token in place $p_{ready}$;
- the pump is not in use - there is a token in place $p_{pump}$;
- *and* the product tank is empty - there is no token in place $p_{tank}$.

Firing transition $t_{drain}$ results in the token distribution seen in Figs. 8.17.a and b.

---

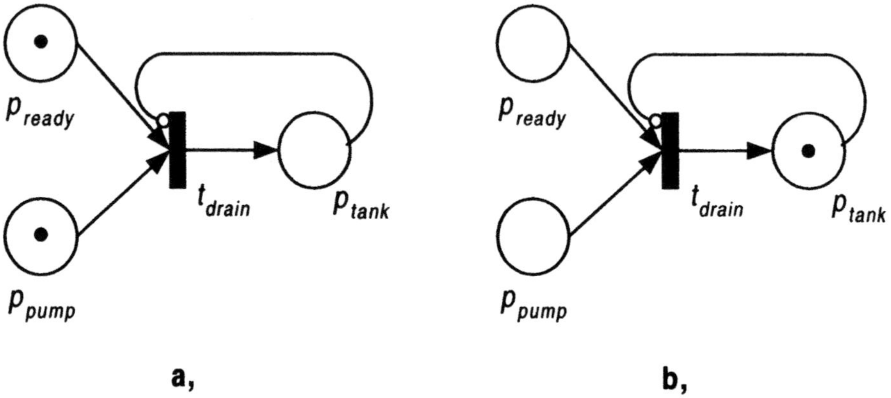


*Figure 8.17.* **Modelling a draining section with inhibitor arcs**

Another application of inhibitor arcs is solving conflict situations between transitions. Directing an inhibitor from the separate input place of one transition to the other transition ensures priority for the first transition over the second as we can see in the following example.

---

**EXAMPLE 8.13** Inhibitor arcs to solve conflict situations

Let us assume we have two identical tanks to store the product in Example 8.11. The Petri net of the draining part is modified as shown in Fig. 8.18.

If both storage tanks are empty then transitions $t_{drain-to-t1}$ and $t_{drain-to-t2}$ are in a conflict situation. We can resolve this conflict by adding an inhibitor arc to the net directing it from place $p_{tank1\_empty}$ to transition $t_{drain-to-t2}$. This modification can be seen in Fig. 8.18. In this case if both tanks are empty - and the other two preconditions are valid, too - then only transition $t_{drain-to-t1}$ will be enabled. The necessary condition for enabling transition $t_{drain-to-t2}$ is to have no tokens in place $p_{tank1\_empty}$.

---

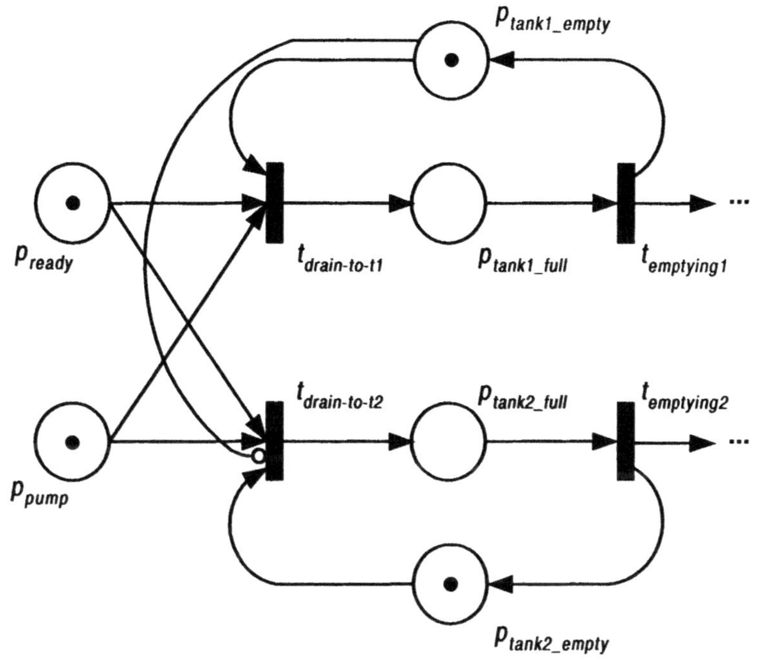


*Figure 8.18.* Modelling of a draining section with two tanks

## 1.3.6 DECOMPOSITION OF PETRI NETS

One of the main advantages of modelling with Petri nets is the capability to describe hierarchical systems. This means that the systems to be modelled can be described on different levels. In the case of a complex system first the models of subprocesses can be made and checked separately then they can be built into the model of the whole system. Both transitions and places can be considered as composite elements, i.e. subnets can be built into them. This process can be repeated arbitrarily.

Another advantage of this method is that in the case of large systems containing a large number of similar subprocesses, the subnets of these elements only have to be made in one instance in advance, then they can be used as modular elements during the modelling process which simplifies the modelling task.

**EXAMPLE 8.14** A simple hierarchical Petri net

If we again consider the reaction part of the simple process system in Example 8.1 then we can use the already developed Petri net model of feeding and reaction to construct a hierarchical model. This model contains a decomposition of a place and a transition as seen in Fig. 8.19.

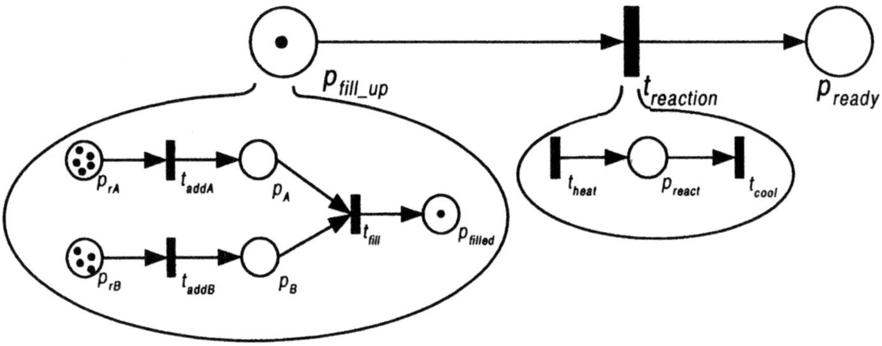


*Figure 8.19.* Decomposition of Petri net elements

### 1.3.7 TIME IN PETRI NETS

As it was shown in the previous examples, time does not appear in an explicit manner in the original Petri net concept. One of the reasons for this is that C. A. Petri developed his tool for modelling communication between serial automata. The firing of transitions (and associated events) is considered to take place in zero time, i.e. instantaneously. This type of transitions is called primitive transitions.

In real discrete event systems, however, there are events which do not take place instantaneously. These are called non-primitive transitions in the model. There are different solutions for handling non-primitive transitions. In a simpler case we can use the decomposition method as it can be seen in Fig. 8.20.

There is a modification of Petri nets in literature, where time is associated explicitly to the firing of a transition. [80]

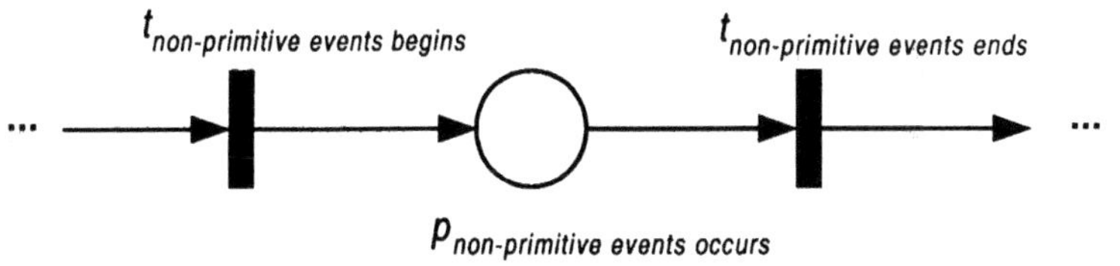


*Figure 8.20.* Decomposition of non-primitive events

## 1.4 THE STATE-SPACE OF PETRI NETS

In the formal definition of Petri nets we introduced marking function $M$, which assigns a non-negative number to each place. This marking value refers to the state of a place and the distribution of tokens refers to the state of the net.

Starting from a given initial distribution of tokens, i.e. from an initial state the enabled transitions can be determined. The firing of these transitions changes the state of their input and output places and so the state of the whole net. In this new state (or in new states) there can be other enabled transitions and their firing changes the net state again. This process is repeated while there is at least one enabled transition in the net.

We can collect the states that resulted from the firing of transitions from a given initial state in the *reachability set.* The formal definition of a reachability set is the following.

**Definition 8.2.** *Let $M_0$ be the initial state in a given Petri net. Denote $R(M_0)$ the set of reachable markings starting from $M_0$, that is the reachability set belonging to $M_0$. Then*

1. $M_0 \in R(M_0)$;

2. *if $M' \in R(M_0)$ and there is transition $t_j$ which is enabled in $M'$ and firing of $t_j$ changes the net state into $M''$ then $M'' \in R(M_0)$.*

---

**EXAMPLE 8.15** A reachability set

Simple analysis shows that the reachability set of the Petri net in Example 8.4 is the following:

$R(M_0) = \{\ (1,0,0),\ (0,1,0),\ (0,0,1)\ \}$

---

## 1.5 THE USE OF PETRI NETS FOR INTELLIGENT CONTROL

Discrete event dynamic system models naturally arise in the following application areas related to intelligent control.

1. Discrete event dynamic system models are traditionally and effectively used in ***design, verification and analysis of operating procedures*** which can be regarded as sequences of external events caused by operator interventions.

2. The ***sceduling and plant–wide control of batch plants*** is another important, popular and rapidly developing field. Batch plants produce a charge of material or a piece of eqiuipment at a time, which we regard as indivisible.

There are various but related approaches to describe a discrete event dynamic system with discrete time and discrete valued variables. These include finite automata, digraph and Petri net models of various kind. From these methods, Petri nets are the most popular and widely used ones.

Most of the approaches to representing such systems use combinatorial or finite techniques. This means that the value of the variables including state, input and output variables, are described by finite sets and the cause-consequence relationships between these discrete states are represented by directed graphs of various kind. This enables to give equivalent representations of a discrete event dynamic system model using various competing techniques in most of the cases.

# 2. THE ANALYSIS OF PETRI NETS

In the previous section we demonstrated the modelling power of Petri nets. Although we used the same process system (see Example 8.1) as an example throughout the whole section, Petri nets can be used for modelling a large variety of systems especially those containing concurrent events.

Modelling a system and the execution of its Petri net model give a lot of information about the basic structure and processes taking place in it but the analysis also ensures provable correct consequences.

In this section first we consider the type of questions that can be raised during an analysis and then introduce two basic analysis methods.

## 2.1 ANALYSIS PROBLEMS FOR PETRI NETS

Petri net properties can be divided into two major classes: *behavioural (or marking dependent) properties* and *structural properties* which are independent of the initial marking, i.e. the initial state. Several properties from both classes are identified and analyzed for Petri nets (see e.g. [78], [79]). In the following we only focus on properties of industrial and practical interest, which are relevant to Petri nets describing controlled discrete event systems.

### 2.1.1 SAFENESS AND BOUNDEDNESS

For a Petri net which is to model a discrete event system, one of the most important questions is boundedness. Boundedness and its special case safeness are related to the limited capacity of places.

A place in a Petri net is bounded if the number of tokens in that place never exceeds a given value. If this maximum value is equal to 1 then the place is called safe.

The interpretation of safeness and boundedness depends on the system to be modelled. In our process system described in Example 8.1 the places representing the states of the reactor or that of the product tank must be safe. The presence of more than one token in these places means that there is a state during the execution of the operating procedure when we want to fill more liquid into the given tank than possible.

The examination of boundedness and safeness can be done for a group of places or for all places in the net. If all places are safe then the net can be called a safe Petri net. If an upper limit $k$ that holds for all places can be determined in the net then the net is a $k$-safe Petri net.

### 2.1.2 CONSERVATION

The *conservation* property is related to the changes in the sum of tokens in a Petri net during execution. A Petri net is *strictly conservative* if the number of tokens is the same in all markings starting from an initial marking.

Strict conservation is a very strong property. It can be useful in the case of modelling resource allocation systems where tokens may represent the resources. It is a very natural requirement for these systems that these tokens are neither created nor destroyed.

It is possible to assign conservation weights to places and check the sum based on the linear combination of tokens computed using the place weights. In this case the *weighted sum* for all reachable markings should be constant for a *conservative* Petri net.

The investigation of conservation can be done for a subset of places, too.

In our process system in Example 8.1 it is useful to investigate this property for the tokens representing pumps.

### 2.1.3 LIVENESS

Liveness addresses the question whether it is always possible to activate a specific transition or the system can reach a state where this transition is "dead". As a generalization of this problem we can investigate whether the system can reach a state where there is no enabled transition at all. This state is called a *dead-lock*. A system can get into a dead-lock when the operating procedure is over but it could also happen that the process stops before the final state. A dead-lock is very dangerous in the latter case because it refers to a state where the operator has no possibility to intervene, that is, the system is out of control.

### 2.1.4 REACHABILITY AND COVERABILITY

During execution different markings can be reached in a Petri net. These markings are either desirable or undesirable from the viewpoint of the operation of the modelled system. The reachability problem addresses the question whether it is possible to reach or avoid a given marking starting from a given initial state.

The coverability problem is the generalization of the reachability problem. Here we investigate whether there is a marking $M''$ in a reachability set of $M_0$ such that $M'' \geq M'$ i.e. $M''$ covers a predefined marking $M'$. (A marking $M''$ covers marking $M'$ if $\forall i : M''_i \geq M'_i$, i.e. each component of marking $M''$ is greater than or equal to the components of marking $M'$.)

The investigation of reachability and coverability can be done for a restricted set of places, too.

It is important to note that the reachability of Petri nets resembles the controllability of LTI continuous time systems (see in Appendix A).

### 2.1.5 STRUCTURAL PROPERTIES

The other possibility in the analysis of Petri nets is the determination of *place and transition invariants*. The place invariant is a set of places, in which the sum of tokens remains constant, independently of which transition fires. Tokens of this set of places are neither generated nor consumed, only "moved" between the places.

The transition invariant is a set of transitions. When these transitions fire starting from an initial state the system returns to the same initial state. Transition invariants correspond to the cyclical behaviour of the modelled system.

## 2.2 ANALYSIS TECHNIQUES

There are two major Petri net analysis techniques: the reachability tree and matrix equations.

The aim of constructing a reachability tree is to answer the initial state dependent questions. It involves the determination of all possible markings that belong to a given initial state. On the other hand matrix equations are used for determining structural properties.

Both techniques can be implemented on a computer. The use of computers is very important during the analysis because apart from some simpler cases the analysis of the above mentioned properties is very difficult without software support.

### 2.2.1 THE REACHABILITY TREE

The reachability tree technique involves the enumeration of all reachable markings from a given initial marking. The method of constructing a reachability tree is the following. Starting from the given initial state as the root of the tree we determine the enabled transitions. The number of enabled transitions in a given state is equal to the number of new markings that will be added as new nodes to the tree. These new nodes are connected to their parent node by directed arcs, which have the colour of the fired transition. We repeat this process for every new node until there is no enabled transition. The terminal nodes of the tree are the "dead" markings where there are no enabled transitions.

It is easy to see that even simple bounded Petri nets can have an infinite reachability tree. To avoid infinite trees we do not perform the investigation of an enabled transition if the new marking is either equal to an earlier one in the tree or it covers another marking which is found on the path leading from the root to this new node.

In the first case, equality, we can mark the new node as a duplicate node. There is then no need to check the enabled transitions and the new markings resulting from the firing of these transitions because it has already been done for the first appearance of this node in the tree.

The second case, when the new marking covers an earlier one lying on the path from the root, refers to the cyclic behaviour of the net. This means that there is a loop of transitions that can be performed an arbitrarily number of times. It is unnecessary to indicate all nodes belonging to each appearance of this loop but somehow we have to refer to them.

In the following we introduce a simple example as an illustration.

---

**EXAMPLE 8.16** Construction of a reachability tree

Let us assume the simple Petri net in Fig. 8.21.a.

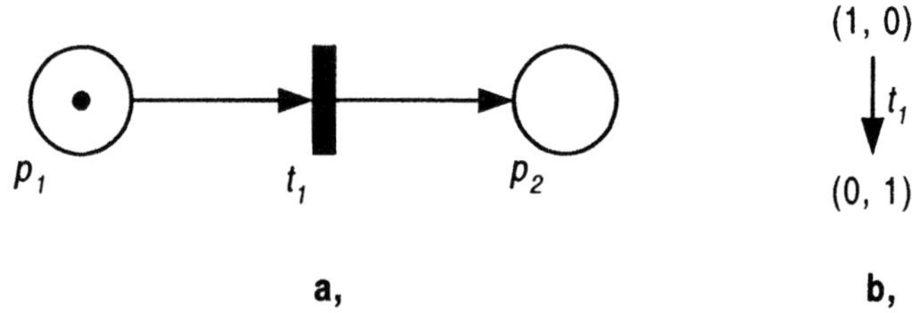


*Figure 8.21.* A simple net and its reachability tree

Starting from the initial state $M_0 = (1,0)$ it is very easy to get to the reachability set of the net: $R(M_0) = \{(1,0),(0,1)\}$. The reachability tree can be seen in Fig. 8.21.b. The last node on the tree refers to a terminal node because there is no enabled transition in this marking.

A slight modification of the net does not change the reachability set but gives an infinite reachability tree (see Fig. 8.22.)

Applying the concept of duplicate nodes, the tree can be reduced to a finite tree (Fig. 8.23).

The last node is a duplicate node, i.e. it refers to a repeated marking in this case.

Let us modify the net again (see Fig. 8.24).

Starting from the initial state $M_0 = (1,0,0)$ both the reachability set and tree will be infinite (see Fig. 8.25).

There are no duplicate nodes in the tree but the comparison of the markings labelled by (*) shows that the third marking covers the initial marking while the fifth covers the previous two labelled.

---

The introduction of symbol $\omega$ can solve the loop indication problem. The symbol $\omega$ represents an arbitrarily large number of tokens. For any constant $a$ the following is true:

$$\omega \pm a = \omega, \qquad a < \omega, \qquad \omega \leq \omega.$$

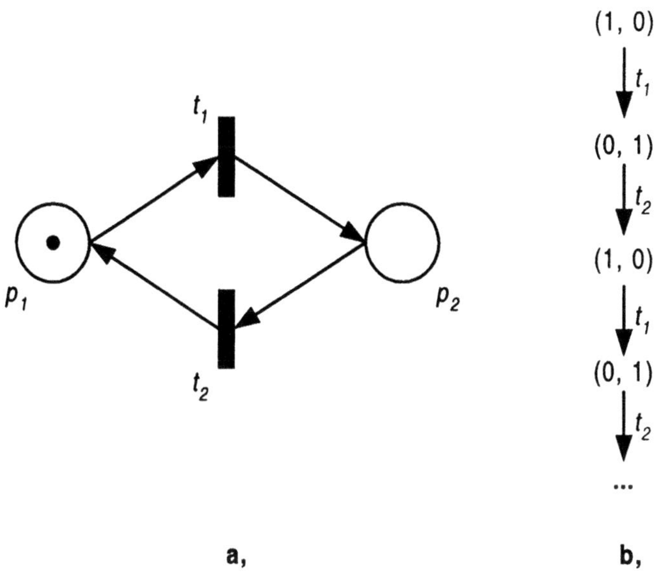


*Figure 8.22.* An infinite reachability tree

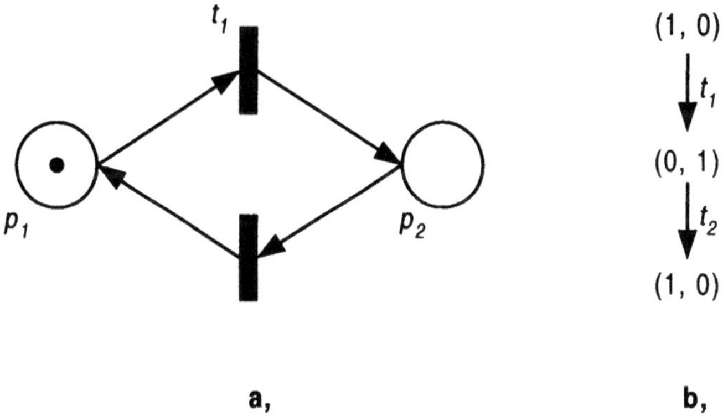


*Figure 8.23.* A reachability tree with duplicate nodes

Applying these modifications and notations, the algorithm for constructing a finite reachability tree is as follows:

1. Let the initial marking be the root of the tree. Let $L$ be the list of new nodes. Add the root to list $L$.

2. If $L$ is empty then the algorithm stops, the reachability tree is ready. Otherwise let $x$ be the first node from the list, let $M_x$ be the marking associated with it and remove $x$ from the list.

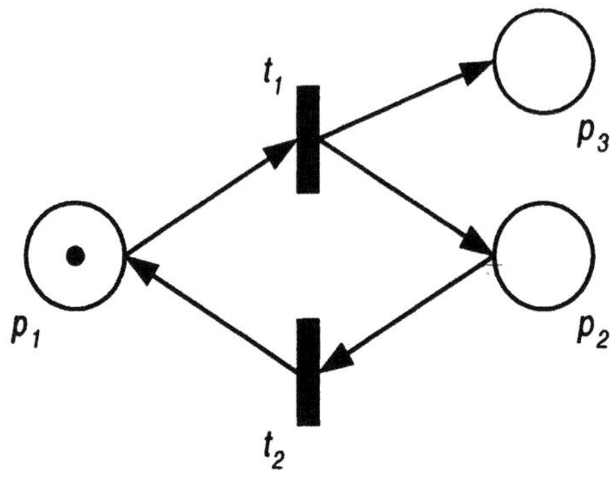


*Figure 8.24.* A modified simple net

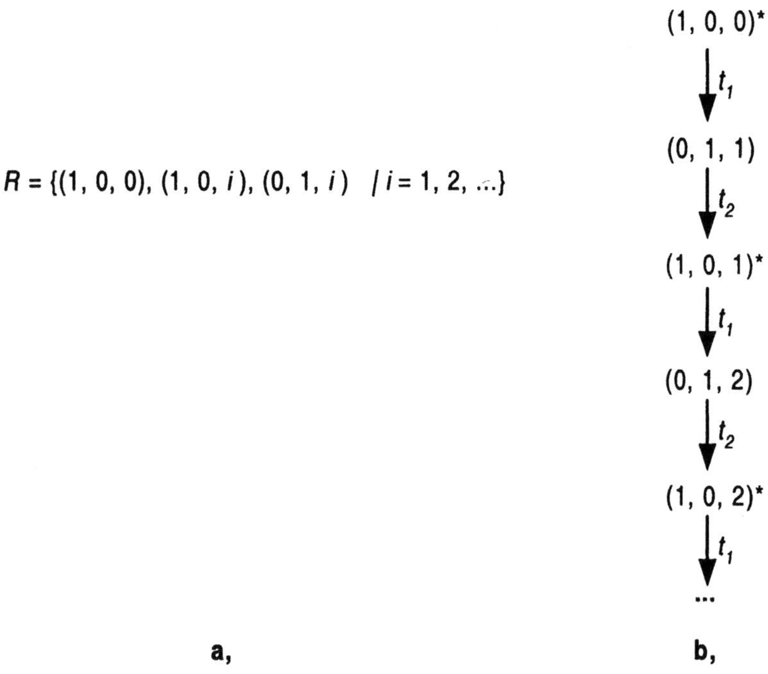


*Figure 8.25.* An infinite reachability set and tree

3. If another node exists in the tree, say $y$, which has the same associated marking with it ($M_x = M_y$) then $x$ is a duplicate node. It will be a terminal node of the tree with the remark *duplication*.

4. If no transition is enabled in marking $M_x$ then this marking is a deadlock in the net. This node will be a terminal node again but with the remark *dead*.

5. For all transitions enabled in marking $M_x$ do the following:

   (a) Create a new node in the reachability tree, connect this node to $x$ with a directed arc labelled by the symbol of the fired transition and add it to the end of list $L$.

   (b) Determine the marking associated with the new node by applying the firing rule. If the parent marking $M_x$ contains the symbol $\omega$ assigned to a place than all of its children markings will contain the symbol $\omega$ in the same place.

   (c) Let us denote the new marking by $M_z$. If there exists a marking say $M_i$ on the path from the root to marking $M_z$ such that $M_z$ covers $M_i$, that is $M_z(p) \geq M_i(p)$ for each place $p$ in the net and there is at least one place, say $p_k$ where $M_z(p_k) > M_i(p_k)$, then the marking $M_z$ will contain the symbol $\omega$ in the co-ordinate referring to the place $p_k$.

6. Go back to Step 2.

---

**EXAMPLE 8.17** Applying of $\omega$

Applying the $\omega$ symbol during the construction of the reachability tree in case of the net in Example 8.16 we get a finite tree as it can be seen in Fig. 8.26.

---

Having constructed the reachability tree of a Petri net, most of the analysis of its properties can be performed by searching in the tree as follows:

- A Petri net is bounded if and only if the symbol $\omega$ does not appear in any of the markings of the tree.

- A Petri net is safe if and only if only zero and one values (0's and 1's) appear in the markings of the tree.

- A transition is dead if and only if it does not appear in the tree.

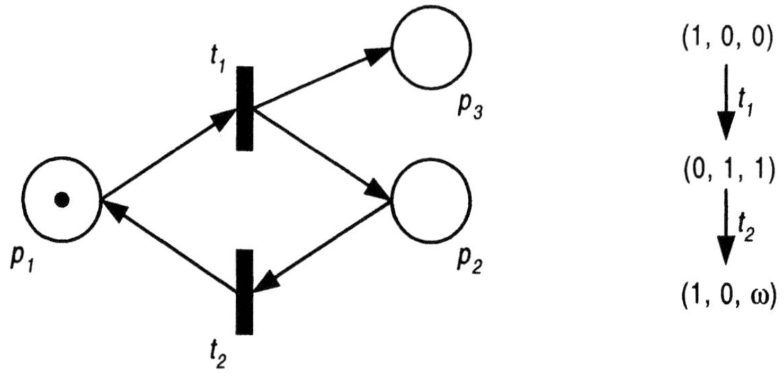


*Figure 8.26.* **Finite reachability tree with using of symbol $\omega$**

- A branch in the tree can refer to transitions either in a concurrent or in a conflict situation. To distinguish the two situations, a deeper analysis of preconditions is needed.

- The reachability and coverability problems can be solved by searching for the predefined marking(s) in the tree.

We have to note that the introduction of symbol $\omega$ causes information loss and as a consequence the liveness of a transition cannot be examined in all cases. This means that the general reachability problem cannot be solved by simply searching in the tree.

The main disadvantage of the analysis with the reachability tree is its exhaustive characteristic. Despite the introduced modifications in the construction, the reachability tree can be very large and this can cause the time and space needed for the construction and search to grow exponentially, therefore this analysis is usually a computationally hard problem.

### 2.2.2 ANALYSIS WITH MATRIX EQUATIONS

An analysis using the reachability graph gives information about the behaviour of the net starting from a given initial state. It would be a great advantage if it could somehow be generalized and we could find a method which solved the analysis problems in a shorter time and in a simpler way than the generation of trees. The invariance analysis can partly give an answer to this request. Using the matrix based description of the Petri net model we can analyze the structural properties of the system.

**The representation of Petri nets by matrices.** Let us represent the Petri net model of the investigated system by an incidence matrix. The first index of an element in the incidence matrices refers to the corresponding place, while the second index refers to the corresponding transition. The number of rows is equal to the number of places and the number of columns is equal to the number of transitions. An entry in the matrix is equal to the difference between the weights of the outcoming and incoming arcs of a transition-place pair:

$$h_{ij} = h_{ij}^{+} - h_{ij}^{-} \tag{8.21}$$

where

$h_{ij}$ an entry of incidence matrix $H = [h_{ij}]$ ;
$h_{ij}^{+} = w(t_i, p_j)$ is the weight of the arc from transition $i$ to place $j$ ;
$h_{ij}^{-} = w(p_j, t_i)$ is the weight of the arc from place $j$ to transition $i$.

---

EXAMPLE 8.18 Representing a Petri net by an incidence matrix

Let us represent the Petri net in Fig. 8.27 by an incidence matrix.

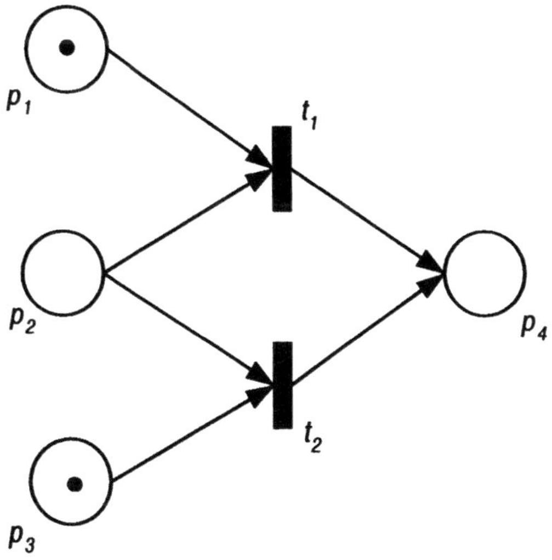


*Figure 8.27.* Simple net

The number of places is equal to 4 while the number of transitions is 2. Hence the incidence matrix $H$ is the following

$$H = \begin{bmatrix} 1 & 0 \\ 1 & 1 \\ 0 & 1 \\ -1 & -1 \end{bmatrix} \qquad (8.22)$$

---

From the point of view of engineering meaning, incidence matrices can be interpreted as follows. An element of an incidence matrix gives the relation between a place and a transition. If an element $h_{ij}$ is not equal to zero then the transition $t_i$ and place $p_j$ are connected. If $h_{ij}$ is a positive number then place $p_j$ is a precondition of transition $t_j$ while if $h_{ij}$ is a negative number then this place is a consequence of it.

A zero entry can have different meanings. It can mean that there is no connection between the given transition and the given place but we get the same entry if the given place is both an input place and an output place of the transition having the same weights. To avoid this information loss we assume that the investigated Petri net is *pure*, i.e. it does not contain any self-loops. If it contains one then we can eliminate it by adding a dummy transition place pair to this self-loop. The column vector of an incidence matrix gives all the preconditions and consequences of a given transition and a row vector defines the connections between a given place and the transitions of the net.

**Determination of the invariants.** The place and transition invariants are the structural properties of a Petri net. The place invariant is a vector of weights. If we multiply this vector by the vector representing the number of tokens in the places, we will get a constant scalar value independently of which transitions fire. The formal definition of the place invariant is as follows.

**Definition 8.3.** *Let us assume that $H$ is the incidence matrix of a Petri net $N =< P, T, F, W >$ and $x \in \mathbb{Q}^{|P|}$ is a $|P|$-dimensional (column) vector of rational numbers. Vector $x$ is defined to be a place invariant if it is a nontrivial solution of the system of linear equations $x^T \cdot H = \underline{0}^T$ where $\underline{0}$ is the zero vector and $^T$ denotes transposition.*

Note that there can be no place invariant to a given Petri net when the equation above does not have any nontrivial solution.

The transition invariant is a set of transitions. When every transition in the invariant fires starting from an initial state the system returns to the same initial state. This leads to the following formal definition.

**Definition 8.4.** *Let us assume a Petri net* $N =< P, T, F, W >$ *above with the incidence matrix* $H$ *and* $y \in \mathbb{Z}^{|T|}$ *is a* $|T|$*-dimensional vector of integer numbers. Vector* $y$ *is defined to be a transition invariant if it is a nontrivial solution of the system of linear equations* $H \cdot y = \underline{0}$.

Here again, a transition invariant to a given Petri net may not exist when the above equation has no nontrivial solution.

We can interpret the invariants from a modelling point of view as follows. Let us assume a model of a resource allocation system that uses Petri nets. In this case certain tokens refer to the resources in the net. If the model works properly a way then the number of these tokens has to be the same in every system state. The places where these tokens can be found during the execution of the net form a place invariant of the system.

The transition invariants correspond to the different cyclical behaviours of the system. Starting from a certain initial state and firing these transitions the system has to return to the same initial state.

# Chapter 9

# FUZZY CONTROL SYSTEMS

Fuzzy control systems are able to describe and handle symbolic as well as uncertain information together with rule-based reasoning [81]-[82].

The sections of this chapter cover the following topics:

- Introduction to fuzziness and to fuzzy control
- The notion of fuzzy sets and the operations on fuzzy sets
- Designing fuzzy rule-based control systems

## 1. INTRODUCTION

Before we turn to the main subject of the chapter we first discuss the notion of fuzzyness and then introduce the notion of fuzzy controllers.

### 1.1 THE NOTION OF FUZZINESS

We can decide whether an element is a member of a set or not by applying the rules of classical set theory. For example, it can be decided whether a car belongs to the products of a given manufacturer. But how can we answer the question 'Is the speed of this car high?' Although the speed of a car can be measured unambiguously but the judgment of fastness depends on the circumstances, too. You could be a fast driver when your speed is only 50 km/h but you are driving in a narrow street packed with parking cars. Similarly, 80 km/h could be slow in a highway where the upper speed limit is 130 km/h.

Let us assume a speed limit of 80 km/h and good driving conditions. Are you a driver obeying the rules if your speed is 79.9 km/h and a fast driver it is 80.1 km/h? Of course it is necessary to draw the line somewhere, but in practice there is a need for a zone of tolerance. It would be better if the maximum speed allowed was defined by taking all the circumstances into account: the slipperiness of the road, the daylight, the condition of the car, the skills of the driver, etc. Even if all these elements are taken into account, the expression 'high speed' could be described by a closed interval rather than a given value. The lower limit of this interval refers to 'not high speed', the upper limit to 'high speed' and the inner elements of the interval refer to more or less high speed. This method does not work for the police but it would be very useful for a car or vehicle driven by a computer.

## 1.2 FUZZY CONTROLLERS

In classical control theory the manipulated variable, i.e. the output of the controller is generally calculated based on the basis of the difference between the reference input and the measured value. All these data are exact numerical values and the calculation is performed by a controller algorithm.

However, it is very natural to formulate rules when describing the operation of a controller instead of an algorithm. These rules are based on experience in most cases and they contain linguistic expressions rather than numerical values. Using the example related to the speed of a car in the previous section we can formulate a rule as follows.

```
If the speed is high and it begins to rain
 then reduce the speed
```

To evaluate this rule the notion of 'high' has to be determined and as we have seen above it can be performed by grading in an interval.

# 2. FUZZY SETS

## 2.1 DEFINITION OF FUZZY SETS

Classical set theory considers the elements of a set as a whole. The elements are often called the members of the set. The universe from which they are selected can be given. It can be decided about every item of the universe whether it belongs to the given set or to its environment, i.e. to the other part of the universe. There is no restriction on the size of the set. There are methods in mathematics to define and handle sets with zero or an infinite number of elements. We usually refer to classical sets as *crisp sets* in fuzzy set theory.

---

**EXAMPLE 9.1** Crisp sets

Let us have the following relation between the input variable $u$ and output variable $y$:

$$y = u^2$$

Assuming that the input can only have positive integer values the results can be given in a tabular form as follows:

| u | 1 | 2 | 3 | 4 | ... |
|---|---|---|---|---|---|
| y | 1 | 4 | 9 | 16 | ... |

Then the set of measurements $M = \{\langle x_i, y_i \rangle, i = 1, 2, \ldots, n\}$ where the measured (output) value is

1. less or equal than 3 contains only one pair of measured values;
2. greater or equal than 16 contains an infinite number of measured value pairs but it is easy to decide whether a given measurement is a member;
3. greater or equal than 5 and less or equal than 8 does not contain a pair, it is an empty set.

These sets can also be defined mathematically:

1. $M_{y \leq 3} = \{\langle u, y \rangle | y \leq 3\}$ e.g. $M_{y \leq 3} = \{\langle 1, 1 \rangle\}$;
2. $M_{y \geq 16} = \{\langle u, y \rangle | y \geq 16\}$ e.g. $M_{y \geq 16} = \{\langle 4, 16 \rangle, \langle 5, 25 \rangle, \ldots\}$;
3. $M_{5 \leq y \leq 8} = \{\langle u, y \rangle | 5 \leq y \leq 8\}$ e.g. $M_{5 \leq y \leq 8} = \emptyset$.

The above sets can be represented in a graphical form, too as it is shown in Fig. 9.1.

---

In the case of finite sets the elements can be listed but it does not work for sets with many or an infinite number of elements. These can be described by means of a predicate and this predicate is evaluated in the universe.

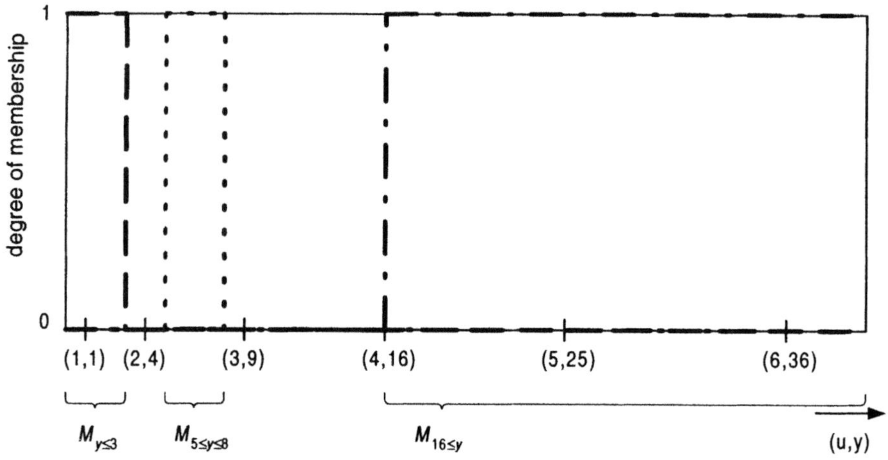


*Figure 9.1.* Crisp sets

Zadeh gave another interpretation of membership [83]. He stated that it was a very hard task to decide whether a given element was part of a set. Repeating the introductory example about fast drivers it is very easy to decide whether one is faster than the maximum speed allowed but it is much harder to define an upper limit taking all circumstances into account. Zadeh proposed to assign a *grade of membership* in the set to each element of the universe. Elements which are obviously members of the set have a grade of membership of 1 while those that definitely do not belong to the set have a 0 grade. Other elements have a grade of membership between 0 and 1 depending on how much they belong to the set. A *membership function* assigns this grade to each element. The concept of membership can be defined in classical set theory, too. In this case the grade of membership is either 0 if the item is not a member of a set, or 1 if it is. In fuzzy set theory classical sets are often called *crisp* sets.

There is no rule about of how to determine the actual value of the grade of membership. It depends on the user's knowledge relating to the behavior or nature of the universe. For example, 100 km/h is a medium high speed in dry weather conditions with good visibility but it is very very high in a thick fog. Membership is often subjective. For a 4 year old kid, a 30 year old man seems very old, while for a 70 year old man he is young.

For fuzzy sets, the concept of universe is similar as it was for classical sets. It contains all the items that can come into consideration but the border between the set and its environment is not given clearly.

---

**EXAMPLE 9.2** Fuzzy sets

Let us consider the same set of measurements as in Example 9.1.

Assuming 6 as a maximum input value we can assign the following membership value $\mu$ to the pairs of sets.

1. to the pairs considered to be high $M_{y>9}$

| u | 1 | 2 | 3 | 4 | 5 | 6 |
|---|---|---|---|---|---|---|
| y | 1 | 4 | 9 | 16 | 25 | 36 |
| $\mu$ | 0 | 0 | 0.2 | 0.6 | 1 | 1 |

2. to the pairs considered to be medium $M_{1<y<25}$

| u | 1 | 2 | 3 | 4 | 5 | 6 |
|---|---|---|---|---|---|---|
| y | 1 | 4 | 9 | 16 | 25 | 36 |
| $\mu$ | 0.1 | 0.4 | 0.9 | 0.8 | 0.1 | 0 |

3. to the pairs considered to be very low $M_{y<3}$

| u | 1 | 2 | 3 | 4 | 5 | 6 |
|---|---|---|---|---|---|---|
| y | 1 | 4 | 9 | 16 | 25 | 36 |
| $\mu$ | 1 | 0.4 | 0 | 0 | 0 | 0 |

4. to the pairs where the measured value $y$ is considered to be much higher than 30 $M_{y>>30}$

| u | 1 | 2 | 3 | 4 | 5 | 6 |
|---|---|---|---|---|---|---|
| y | 1 | 4 | 9 | 16 | 25 | 36 |
| $\mu$ | 0 | 0 | 0 | 0 | 0 | 0.5 |

The graphical representation of these values can be seen in the graph in Fig. 9.2.

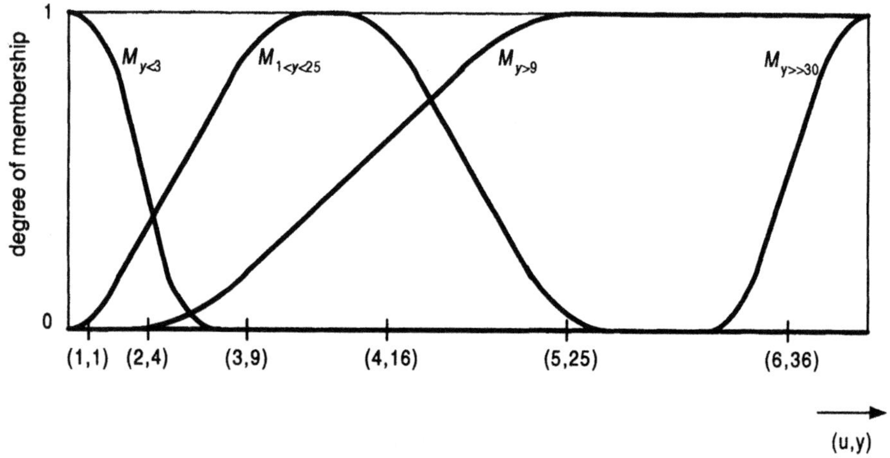


*Figure 9.2.* Fuzzy sets

---

In the above example we assign a grade of membership to each element of the universe. This grade varies between 0 and 1. Elements with a non-zero grade form the *support* of the fuzzy set.

It is not necessary to assign the maximum grade value to an item of a set as we can see in the fourth case. We refer to a fuzzy set as *normalized* if the maximum grade value is equal to one. Normalization can be easily done by dividing each membership value by the maximum value.

In the case of fuzzy sets we often use *linguistic variables* to describe membership criteria. The expressions '*high*', '*medium*' or '*low*' and others are useful terms for the definition of fuzzy sets.

Depending on the nature of the universe a membership function can be represented either in a continuous or in a discrete form. For continuous representation several types of membership functions can be defined. The most important ones are

- *bell-shaped curves*, which are based on exponential functions like the standard Gaussian distribution function with a maximum value of 1

$$\mu(x) = e^{-\frac{(x-x_0)^2}{2\sigma^2}} \tag{9.1}$$

where $x$ is the independent variable on the universe, $x_0$ is the position of the peak relative to the universe and $\sigma$ is the standard deviation; or other types of exponential functions, for example

$$\mu(x) = 1 - e^{-\left(\frac{\sigma}{x_0 - x}\right)^a} \tag{9.2}$$

where $a$ controls the gradient of sloping slides.

- *s-curves*, which are based on the cosine function

$$s(x_0, b, x) = \begin{cases} 0, & x < x_0 - b \\ \frac{1}{2} + \frac{1}{2}cos(\frac{x - x_0}{b}\pi), & x_0 - b \leq x \leq x_0 \\ 1, & x > x_0 \end{cases}$$

where $b$ is the width of the sloping section and $x_0$ is the coordinate of the peak.

- *z-curves* or *decline s-curves*, which are the reflections of $s$-curves. i.e.

$$z(x_0, b, x) = \begin{cases} 0, & x < x_0 \\ \frac{1}{2} + \frac{1}{2}cos(\frac{x - x_0}{b}\pi), & x_0 \leq x \leq x_0 + b \\ 1, & x > x_0 + b \end{cases}$$

- *$\pi$-curves*, which are the combination of $s$- and $z$-curves, such that there is a flat interval rather then a peak near the maximum membership value:

$$\pi(x_1, x_2, b, x) = min(s(x_1, b, x), z(x_2, b, x)) \tag{9.3}$$

- *linear representations* like simple straight lines either increasing or decreasing

$$\ell_{/}(x_\ell, x_r, x) = \begin{cases} 0, & x < x_\ell \\ \frac{x - x_\ell}{x_r - x_\ell}, & x_\ell \leq x \leq x_r \\ 1, & x > x_r \end{cases}$$

and triangular shape curves

$$\ell_{\wedge}(x_\ell, x_c, x_r, x) = \begin{cases} 0, & x < x_\ell \\ \frac{x - x_\ell}{x_c - x_\ell}, & x_\ell \leq x \leq x_c \\ 1 - \frac{x - x_c}{x_r - x_c}, & x_c \leq x \leq x_r \\ 0, & x > x_r \end{cases}$$

If $x_r \neq 1$ in the case of increasing straight lines or $x_\ell \neq 1$ for decreasing lines then these are called *shouldered* curves or fuzzy sets.

- *irregularly shaped* and *arbitrary curves*
  There are some cases when the curves mentioned cannot properly describe the changes in membership value. Let the universe be the age of drivers and let the membership function describe the risk of driving at high speed as an example. The resulting curve has its maximum points at younger and very old ages while minimum at middle ages.

- *discrete representation of fuzzy sets*
  In some cases it is more convenient to represent continuous sets in discrete forms. For this, we pick a given number of points from the universe in an equidistant manner and insert them into functions listed above. The result is a corresponding list of membership values.

  Discrete fuzzy sets can be arrived at if we simply list the elements from the universe with their membership values. These data can be taken from experimental observations.

The graphical representation of most of the curves listed above can be seen in Figs. 9.3-9.5.

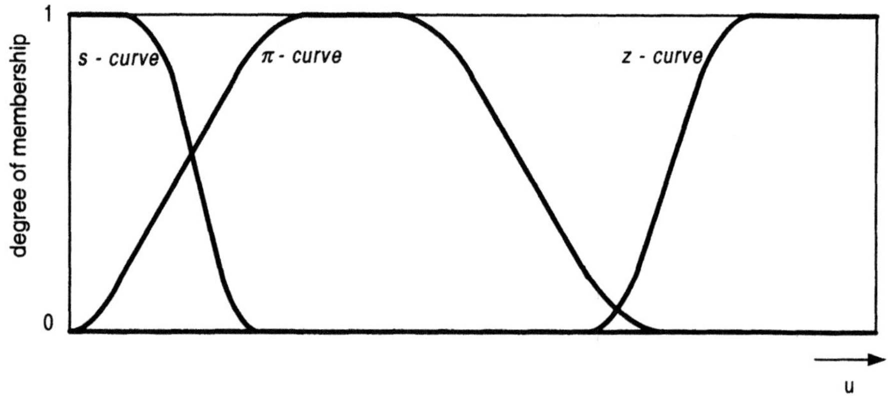


*Figure 9.3.* **s-, π- and z-curves**

For a universe with discrete items, the membership function is implemented as a vector of discrete values. In this case, we can substitute

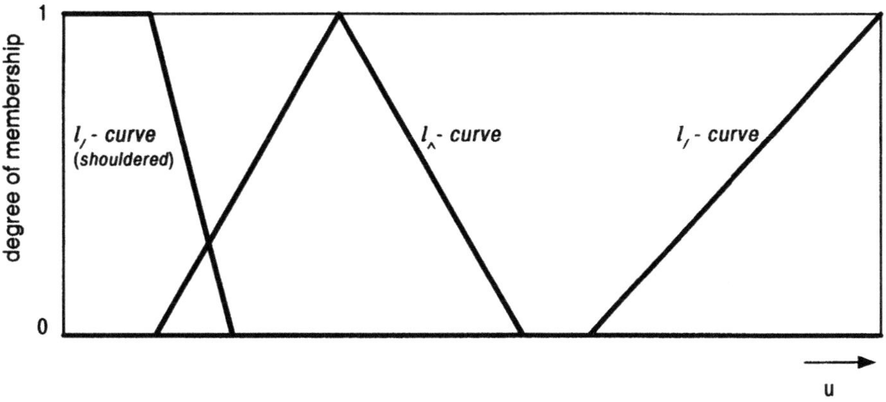


*Figure 9.4.* **Linear representations**

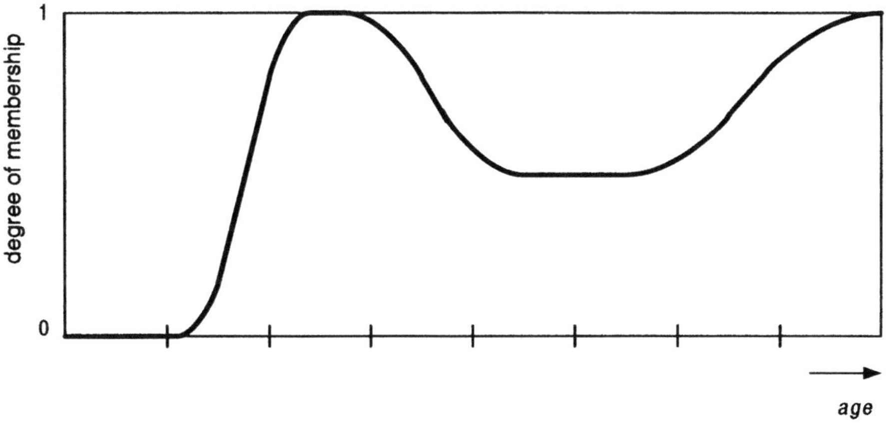


*Figure 9.5.* **Irregular curves**

the discrete input data into the appropriate membership function and calculate membership values.

Summarizing the notion of fuzzy sets we can state that a fuzzy set $A$ is a set of ordered pairs over the universe $U$

$$A = \{\langle x, \mu(x) \rangle\} \tag{9.4}$$

where $x \in U$ and $\mu(x)$ is its grade of membership in $A$. An item $x$ can be either a scalar or a vector variable depending on the nature of the underlying universe. The pair $\langle x, \mu(x) \rangle$ is a fuzzy *singleton*.

According to the Eq. (9.4) a fuzzy set can be considered as a union of fuzzy singletons, especially in the case of discrete representation. Assume a fuzzy set with $n$ elements. Its formal definition is then as follows:

$$A = \{\langle x, \mu(x_1)\rangle, \langle x, \mu(x_2)\rangle, \dots, \langle x, \mu(x_n)\rangle\}.$$

However, it is more convenient to refer to a fuzzy set as a vector of membership function values,

$$a = \mu_{\mathbf{a}} = [\mu(x_1) \;\; \mu(x_2) \;\; \dots \;\; \mu(x_n)]^T$$

omitting the universe. In the next examples of this chapter this latter notation will be used.

There is a distinction between a fuzzy membership function and a probability distribution function in the sense of mathematical statistics. Returning to the 'driving fast' problem the probability function gives the most probable speed of the observed cars, say 85 km/h while the membership function of the *fast drivers* fuzzy set assigns 1 either to the speed 100 km/h or 150 km/h although the probability of the latter is low. The fuzzy membership function determines the possibility of an event. In general we can say that if an event is highly probable it must also be possible but a possible event is not necessarily highly probable.

## 2.2 OPERATIONS ON FUZZY SETS

There are well-known set operations in classical set theory. If $A = \{1, 2, \dots, 10\}$ and $B = \{10, 20, \dots, 100\}$ are two crisp sets then

- the union of the two sets is

$$A \cup B = \{1, \dots, 9, 10, \dots, 100\};$$

- the intersection of the two sets is

$$A \cap B = \{10\};$$

- the complement of set $A$ is

$$\neg A = \{11, 12, \dots, \infty\}$$

provided we have positive integers as our universe.

### 2.2.1 PRIMITIVE FUZZY SET OPERATIONS

We have seen that the membership function plays a specific role in the case of fuzzy sets because it gives the grade of membership in the set. Zadeh defined the fuzzy set operators on the basis of their impact on the membership function [83], [84], [85].

There are three ***primitive fuzzy set operations*** as follows. Let $A = \{\langle x, \mu_A(x)\rangle\}$ and $B = \{\langle y, \mu_B(y)\rangle\}$ be two fuzzy sets over the same universe $U$. Then

- the union of the two sets is

$$A \cup B = A \textbf{ or } B \triangleq A \textbf{ max } B;$$

where **max** is an item-by-item maximum operation between corresponding membership values of $A$ and $B$:

$$\mu_{A \cup B}(x) = \textbf{max}(\mu_A(x), \mu_B(x)) \quad \text{for all } x \in U;$$

- the intersection of the two sets is

$$A \cap B = A \textbf{ and } B \triangleq A \textbf{ min } B;$$

where **min** is an item-by-item minimum operation between corresponding membership values of $A$ and $B$:

$$\mu_{A \cap B}(x) = \textbf{min}(\mu_A(x), \mu_B(x)) \quad \text{for all } x \in U;$$

- the complement of set $A$ is

$$\neg A = \textbf{not} A \triangleq 1 - A.$$

where each membership value of $A$ is substracted from 1:

$$\mu_{\neg A}(x) = 1 - \mu_A(x) \quad \text{for all } x \in U.$$

Assume the discrete valued membership functions $\mu_A$ and $\mu_B$ with $\mu_A$, $\mu_B \in \{0, 0.25, 0.50, 0.75, 1.00\}$. The truth tables of the fuzzy **or** and **and** operations are as follows:

| **or** | 0.00 | 0.25 | 0.50 | 0.75 | 1.00 |
|---|---|---|---|---|---|
| 0.00 | 0.00 | 0.25 | 0.50 | 0.75 | 1.00 |
| 0.25 | 0.25 | 0.25 | 0.50 | 0.75 | 1.00 |
| 0.50 | 0.50 | 0.50 | 0.50 | 0.75 | 1.00 |
| 0.75 | 0.75 | 0.75 | 0.75 | 0.75 | 1.00 |
| 1.00 | 1.00 | 1.00 | 1.00 | 1.00 | 1.00 |

| **and** | 0.00 | 0.25 | 0.50 | 0.75 | 1.00 |
|---|---|---|---|---|---|
| 0.00 | 0.00 | 0.00 | 0.00 | 0.00 | 0.00 |
| 0.25 | 0.00 | 0.25 | 0.25 | 0.25 | 0.25 |
| 0.50 | 0.00 | 0.25 | 0.50 | 0.50 | 0.50 |
| 0.75 | 0.00 | 0.25 | 0.50 | 0.75 | 0.75 |
| 1.00 | 0.00 | 0.25 | 0.50 | 0.75 | 1.00 |

The effect of these operators is demonstrated in the following example.

---

**EXAMPLE 9.3** Fuzzy set operators

Let the universe $U$ be the set of cars characterized by their cylinder capacity in liters: $U = \{1.0, 1.2, 1.4, 1.6, 1.8, 2.0\}$. Let us assume that the acceleration and consumption of a car only depends on cylinder capacity. Then the fuzzy set *low consumption* ($LC$) may be defined as

| $U$ | 1.0 | 1.2 | 1.4 | 1.6 | 1.8 | 2.0 |
|---|---|---|---|---|---|---|
| $\mu_{LC}$ | 1.0 | 0.9 | 0.7 | 0.5 | 0.2 | 0.0 |

and the fuzzy set *high acceleration* ($HA$) is

| $U$ | 1.0 | 1.2 | 1.4 | 1.6 | 1.8 | 2.0 |
|---|---|---|---|---|---|---|
| $\mu_{HA}$ | 0.0 | 0.1 | 0.4 | 0.5 | 0.8 | 1.0 |

If we want to buy a car with *low consumption* **and** *high acceleration* then the intersection of these fuzzy sets should be computed as $\mu_{LC\cap HA} = \mathbf{min}(\mu_{LC}, \mu_{HA})$:

| $U$ | 1.0 | 1.2 | 1.4 | 1.6 | 1.8 | 2.0 |
|---|---|---|---|---|---|---|
| $\mu_{LC}$ | 1.0 | 0.9 | 0.7 | 0.5 | 0.2 | 0.0 |
| $\mu_{HA}$ | 0.0 | 0.1 | 0.4 | 0.5 | 0.8 | 1.0 |
| $\mathbf{min}(\mu_{LC}, \mu_{HA})$ | 0.0 | 0.1 | 0.4 | 0.5 | 0.2 | 0.0 |

But if we need a car with *low consumption* **or** *high acceleration* then we need the union of these fuzzy sets, $\mu_{LC\cup HA} = \mathbf{max}(\mu_{LC}, \mu_{HA})$:

| $U$ | 1.0 | 1.2 | 1.4 | 1.6 | 1.8 | 2.0 |
|---|---|---|---|---|---|---|
| $\mu_{LC}$ | 1.0 | 0.9 | 0.7 | 0.5 | 0.2 | 0.0 |
| $\mu_{HA}$ | 0.0 | 0.1 | 0.4 | 0.5 | 0.8 | 1.0 |
| $\mathbf{max}(\mu_{LC}, \mu_{HA})$ | 1.0 | 0.9 | 0.7 | 0.5 | 0.8 | 1.0 |

The set of cars with **not** *low consumption* is the complement of the fuzzy set $LC$

| $U$ | 1.0 | 1.2 | 1.4 | 1.6 | 1.8 | 2.0 |
|---|---|---|---|---|---|---|
| $\mu_{LC}$ | 1.0 | 0.9 | 0.7 | 0.5 | 0.2 | 0.0 |
| $\mu_{\neg LC}$ | 0.0 | 0.1 | 0.3 | 0.5 | 0.8 | 1.0 |

Assuming s- and z-curves for these membership functions, the results are shown in Figs. 9.6-9.8.

---

Similarly to the case of logical operations (see section 2.1 in Chapter 2) commutativity, associativity, distributivity, DeMorgan rules, absorption and idempotency are valid in the case of fuzzy operations **and** and **or** but exclusion is not satisfied:

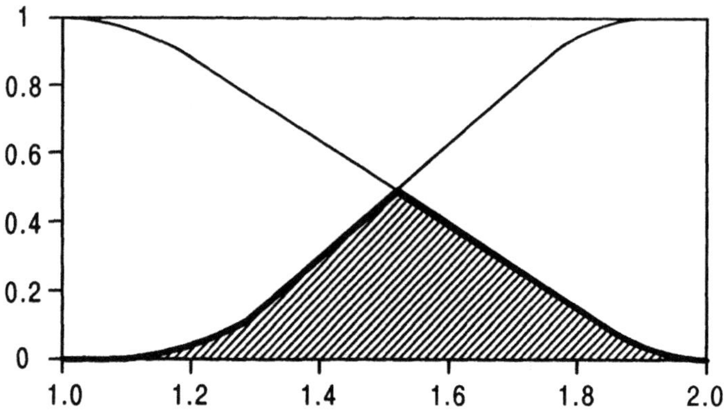


*Figure 9.6.* The fuzzy AND operator

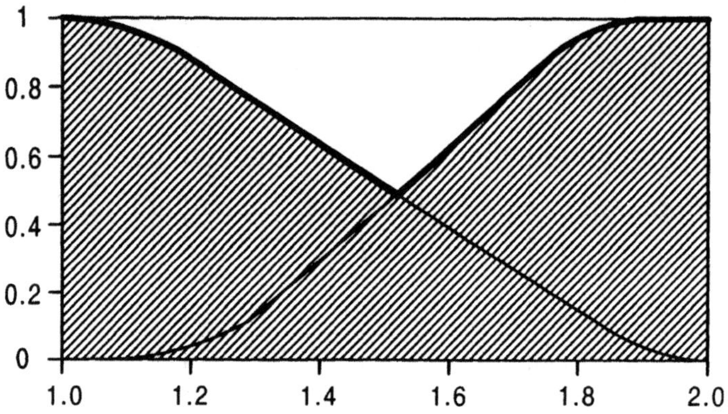


*Figure 9.7.* The fuzzy OR operator

| | |
|---|---|
| commutativity | $a$ **or** $b = b$ **or** $a$ |
| | $a$ **and** $b = b$ **and** $a$ |
| associativity | $(a$ **or** $b)$ **or** $c = a$ **or** $(b$ **or** $c)$ |
| | $(a$ **and** $b)$ **and** $c = a$ **and** $(b$ **and** $c)$ |
| distributivity | $a$ **or** $(b$ **and** $c) = (a$ **or** $b)$ **and** $(b$ **or** $c)$ |
| | $a$ **and** $(b$ **or** $c) = (a$ **and** $b)$ **or** $(b$ **and** $c)$ |
| DeMorgan | **not** $(a$ **and** $b) = ($ **not** $a)$ **or** $($ **not** $b)$ |
| | **not** $(a$ **or** $b) = ($ **not** $a)$ **and** $($ **not** $b)$ |
| absorption | $(a$ **and** $b)$ **or** $a = a$ |
| | $(a$ **or** $b)$ **and** $a = a$ |
| idempotency | $a$ **or** $a = a$ |
| | $a$ **and** $a = a$ |
| exclusion | $a$ **or** $\neg a \neq 1$ |
| (not satisfied) | $a$ **and** $\neg a \neq \emptyset$ |

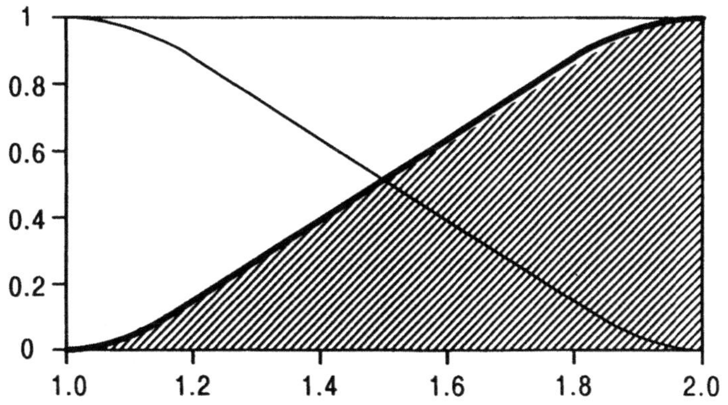


*Figure 9.8.* The fuzzy NOT operator

---

**EXAMPLE 9.4** Example 9.3 cont.

The fuzzy set of cars with *low consumption* **and not** *low consumption* is

| $U$ | 1.0 | 1.2 | 1.4 | 1.6 | 1.8 | 2.0 |
|---|---|---|---|---|---|---|
| $\mu_{LC\cap\neg LC}$ | 0.0 | 0.1 | 0.3 | 0.5 | 0.2 | 0.0 |

and the cars with *low consumption* **or not** *low consumption* is

| $U$ | 1.0 | 1.2 | 1.4 | 1.6 | 1.8 | 2.0 |
|---|---|---|---|---|---|---|
| $\mu_{LC\cup\neg LC}$ | 1.0 | 0.9 | 0.7 | 0.5 | 0.8 | 1.0 |

---

One can find several other fuzzy operators which are based on the extension of the operations **or** and **and** through relatively simple algebraic transformations defined in literature [81].

### 2.2.2 LINGUISTIC MODIFIERS

As it was mentioned earlier we can use linguistic variables, such as **high**, **medium** or **low** for the definition of fuzzy sets. Similarly to spoken language we can add *linguistic modifiers* to these variables to extend or narrow their meaning. The most important groups of linguistic modifiers and their effects are summarized in the following.

- *Approximation of Fuzzy Sets*
  The *approximation* modifiers convert a scalar value into a fuzzy set with a bell-shaped membership function or modify the 'base' of an existing bell-shaped fuzzy set. The most common approximation modifiers are **about**, **around**, **near** and **close to**.

- *Restriction of Fuzzy Sets*
  There are two modifiers, **below** and **above**, which can be used for modifying the shape of linear or bell-shaped membership functions. The modifier **below** can be used if the membership function increases as the universe moves from left to right, while for the applicability of **above** the declination of the membership function is needed.

- *Intensification and Dilution of Fuzzy Sets*
  The *intensification* modifiers **very** and **extremely** (or **very very**) and *dilution* modifiers as **somewhat** (or **morl**), and **greatly** are the most frequently used modifiers. The intensification modifiers can be given in the following form

  $$\textbf{int}\ \mu(x) = \mu^n(x) \tag{9.5}$$

  where **int** refers to an intensification modifier and $n \geq 2$. The value of $n$ is 2 in the case of the modifier **very** and 3 for **extremely**.

  Dilution modifiers have a similar definition equation, except that the power

  $$\textbf{dil}\ \mu(x) = \mu^{1/n}(x). \tag{9.6}$$

  The value of $n$ is 2 in the case of the modifier **somewhat** and 1.4 for **greatly**.

  These modifiers have an interesting property: they can be combined and their combination is commutative.

Example 9.5 shows the effect of these modifiers.

---

**EXAMPLE 9.5** Linguistic modifiers

*Modifier* **about**
Let us assume an operating procedure containing the step:

`'Keep the controlled variable` **about** `50 °C'.`

This instruction defines a fuzzy set with a bell shaped membership function where the central value is 50 centigrade. The graphic representation of this fuzzy set can be seen in Fig. 9.9.

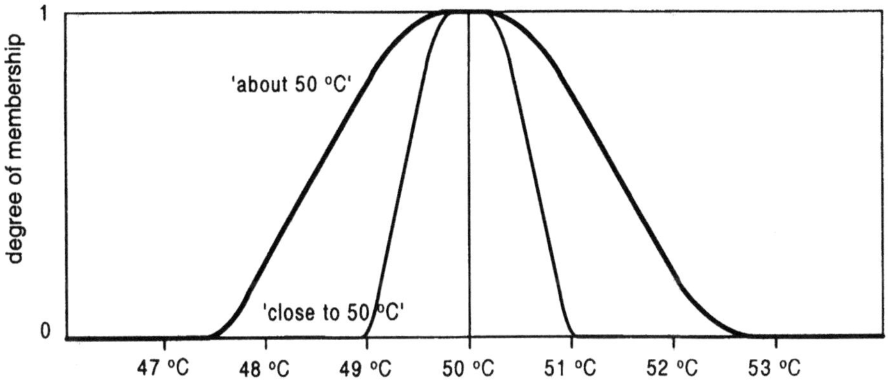


*Figure 9.9.* Linguistic modifiers **about** and **close to**

*Modifier* **below**
As the next case assume a step:

`'Keep the controlled variable` **below** `50 °C'.`

If there is no other constraint then the resulting fuzzy set can be seen in Fig. 9.10.

*Modifiers* **very** and **somewhat**
Assume the fuzzy set *high temperature* with linear representation in Fig. 9.11. The effect of modifiers **very** and **somewhat** is shown in the Fig. 9.11. Obviously, the fuzzy set *very high temperature* refers to a higher temperature zone, i.e. the modifier **very** narrows the original fuzzy set. On the other hand, the modifier **somewhat** makes the original expression *high temperature* more uncertain and it results in a wider fuzzy set.

*Combination of Modifiers*
Using the modifiers **very** and **below** we can form the fuzzy set **very below** 50°C, which refers to the operating step:

`'Keep the controlled variable` **very below** `50 centigrade.'`

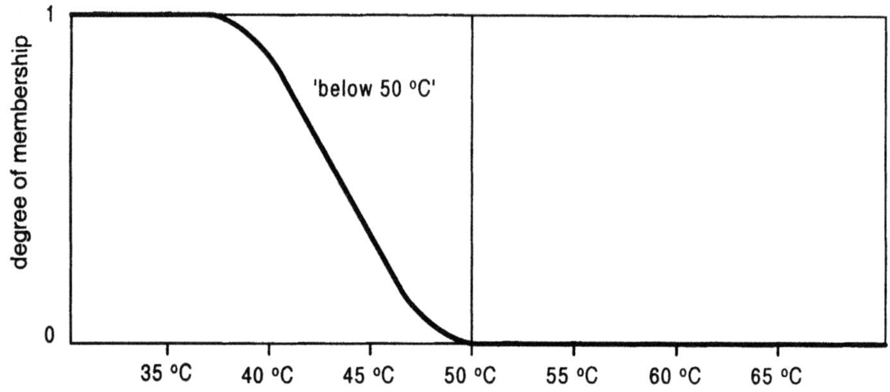


*Figure 9.10.* Linguistic modifier **below**

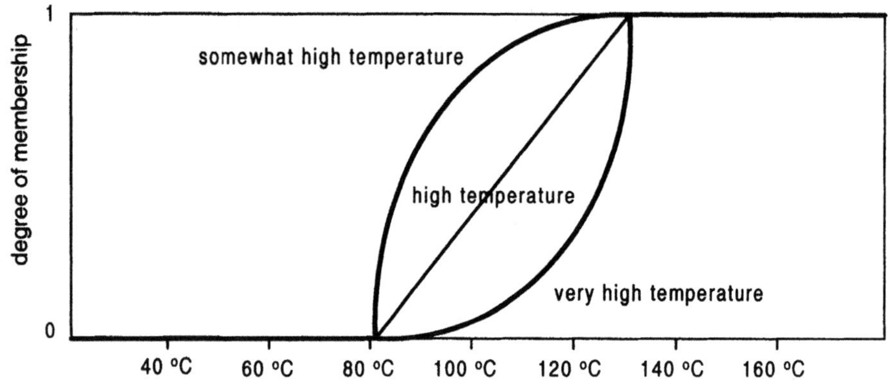


*Figure 9.11.* Linguistic modifiers **very** and **somewhat**

The resulting fuzzy set is in Fig. 9.12.

---

## 2.3 INFERENCE ON FUZZY SETS

As it was mentioned in the introduction, fuzzy controllers contain '**if** – **then**' type rules describing their operations. The conditional part of a rule consists of one or more statements and its application depends on the result of their evaluation. In the case of fuzzy controllers these statements are fuzzy sets and the performed action depends on the value

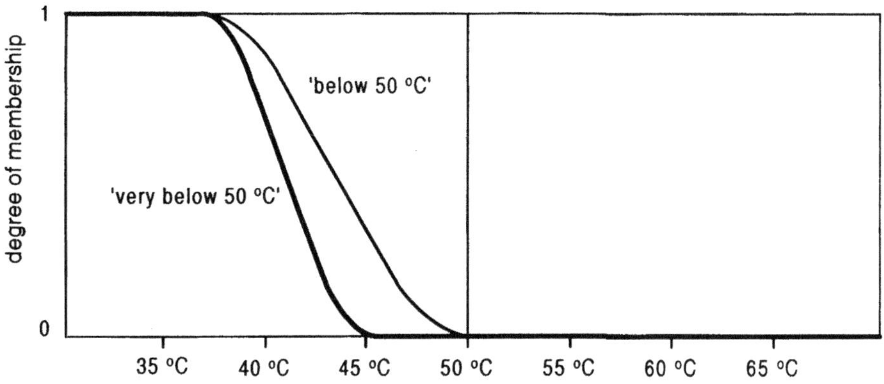


*Figure 9.12.* A combination of linguistic modifiers **very** and **below**

of the membership functions. The conditional part contains at least two terms, i.e. two fuzzy sets in general, and we have to define the relation between these sets. In simple cases these relations contain elements belonging to the same universe but there can be relations between different fuzzy sets defined on different universes.

In this section, we will first deal with the problem of composing relations between fuzzy sets then with the method of inference.

### 2.3.1 RELATION BETWEEN FUZZY SETS

In most cases we want to infer another fact(s) from a fact we find no direct relationship between them. But there can be other facts what we can use as 'transmitters', i.e. we can conclude to these facts from the initial fact and from them to the goal fact. In the case of fuzzy logic there is no ambiguous evidence for truth of a fact so the inference from one fact to another can be characterized by a given degree of possibility as we see in the following example.

---

**EXAMPLE 9.6** Relation

Let us have three universes $P, Q$ and $S$. In the universe $P$ and $Q$ there is only one element $p$ and $q$ respectively, while $S$ has two elements $s_1, s_2$. Assume the elements of $P$ and $Q$ are events while elements of $S$ are states. Let us define a *fuzzy relation* (or shortly relation) $\mathbf{R}_1$ between $P$ and $S$ with the meaning 'an event $a$ causes a state $b$ in a given degree', and a relation $\mathbf{R}_2$ between $Q$ and $S$ with the meaning 'a state $b$ is a precondition of event $a$ in a given degree'.

Fuzzy relations are given in a table containing the degrees of possibility between the elements of the universes being in the relation. This way of specification resembles the definition of a fuzzy set where the values of the membership function over the universe are also given in the form of a table.

The relation between $P$ and $S$ is as follows.

| $\mathbf{R}_1$ | $s_1$ | $s_2$ |
|---|---|---|
| $p$ | 0.3 | 0.9 |

And the relation between $Q$ and $S$ is

| $\mathbf{R}_2$ | $q$ |
|---|---|
| $s_1$ | 0.9 |
| $s_2$ | 0.7 |

We can conclude the following statements from the tables

`( event` $p$ `causes the state` $s_1$ `in degree 0.3`

**and**

`state` $s_1$ `is a precondition of event` $q$ `in degree 0.9 )`

`( event` $p$ `causes the state` $s_2$ `in degree 0.9`

**and**

`state` $s_2$ `is a precondition of event` $q$ `in degree 0.7 )`

From the first statement we can conclude that event $p$ generates event $q$ in degree 0.3 because there is a logical connection **and** between the first and the second part of the logical sentence. Similarly, it follows from the second that $p$ generates $q$ in degree 0.7. Formulate these two sentences as one logical sentence and we get

`( event` $p$ `generates event` $q$ `in degree 0.3`

**or**

`event` $p$ `generate event` $q$ `in degree 0.7 )`

Now there is a connection **or** between the two parts which requires to compute the maximum of the degrees, and it results in the following conclusion

`Event` $p$ `generates event` $q$ `in degree 0.7`

This example contains relations between two fuzzy sets. In the following we formally define binary relations. These can easily be generalized for arbitrarily number of sets.

**Definition 9.1. Composition of binary fuzzy relation**
*Given two fuzzy sets both in matrix form. Their composition is*

$$W = U \circ V \tag{9.7}$$

*where* $\circ$ *is an* **inner or – and product**.

*The* **inner or-and product** *or* **max-min composition** *defined above is a binary relation between two fuzzy sets, which is a fuzzy subset of the Cartesian product of their universes.*

*Assume the fuzzy sets are represented in matrix form and for the definition it is necessary that in the relation the matrix of the first member has the same number of columns as the rows of the matrix of the second member. The defined operation is very similar to the ordinary matrix product except that we apply the operator* **and** *instead of multiplication and the* **or** *instead of summation. Using logical operators* $\wedge$ *and* $\vee$ *rather than operators* **and** *and* **or** *respectively, the result of inner product can be given in the following form:*

$$w_{ij} = (u_{i1} \vee v_{1j}) \wedge (u_{i2} \vee v_{2j}) \wedge \ldots \wedge (u_{ip} \vee v_{pj}) = \bigvee_{k=1}^{p} u_{ik} \wedge v_{kj} \tag{9.8}$$

The defining equation (9.8) of **inner or-and product** explains the other name, **max-min composition** if we recall (see section 2.2.1 in this chapter) that **and** is computed by taking the minimum and **or** is by taking the maximum of the degrees of possibility.

It is interesting to note that the max-min composition is distributive for **or** but not for **and**.

### 2.3.2 IMPLICATION BETWEEN FUZZY SETS

As we have seen before in Chapter 2, rules can be described using the implication operation. Implication is a logical operation and it has the following standard form

$$P \longrightarrow Q \ . \tag{9.9}$$

It can be read as $P$ *implies* $Q$, where $P$ and $Q$ are facts or events of the investigated system. The truth table of the implication can be found in section 2.1 of Chapter 2.

But how does the implication work in the case of fuzzy sets? Let us try it in the following example.

**EXAMPLE 9.7** Implication on fuzzy sets

Let $e$ be the error signal by $e$ and $u$ the controlled input variable (control signal) in a closed loop controlled system. Define the set

$$U_e = \{-5,\ -2.5,\ 0,\ 2.5,\ 5\}$$

as the universe of $e$ and

$$U_u = \{-2,\ 0,\ 2\}$$

for $u$, both in voltage range. Assume there are fuzzy sets for both $e$ and $u$ in the following form

| | |
|---|---|
| a large positive error is | $\mu_{\mathbf{lpe}} = [0\ 0\ 0\ 0.6\ 1]^T$ |
| a small positive error is | $\mu_{\mathbf{spe}} = [0\ 0\ 0.3\ 1\ 0.3]^T$ |
| a zero error is | $\mu_{\mathbf{ze}} = [0\ 0.3\ 1\ 0.3\ 0]^T$ |
| a small negative error is | $\mu_{\mathbf{sne}} = [0.3\ 1\ 0.3\ 0\ 0]^T$ |
| a large negative error is | $\mu_{\mathbf{lne}} = [1\ 0.6\ 0\ 0\ 0]^T$ |

| | |
|---|---|
| a positive control signal is | $\mu_{\mathbf{pcs}} = [0\ 0.2\ 1]^T$ |
| a zero control signal is | $\mu_{\mathbf{zcs}} = [0.1\ 1\ 0.1]^T$ |
| a negative control signal is | $\mu_{\mathbf{ncs}} = [1\ 0.2\ 1]^T$. |

Let a simple control rule be:

$$e \longrightarrow u \tag{9.10}$$

If the actual value of the error signal is equal to 10, then the error is regarded as a "large positive error". We then use the fuzzy set **lpe** and we can conclude that the error signal 10 implies the positive control signal in a degree of 1 and it also implies the zero control signal but only in a degree of 0.2 and the negative control signal in a degree of 0.

At the same time the error signal 5 implies the positive control signal in the degree of 0.6, the zero control signal in a degree of 0.1 and the negative control signal in a degree of 0. The other three control signals have a zero value in the fuzzy set **lpe** so they have no impact on the control signal.

Based on this example, the definition of fuzzy implication is as follows [85].

**Definition 9.2. Implication on fuzzy sets**

*Let $A$ and $B$ be two fuzzy sets, not necessarily on the same universe. The implication between the two fuzzy sets is the following operation*

$$A \longrightarrow B \quad \triangleq \quad A \times B \tag{9.11}$$

*where $\times$ is an* outer *product of the matrices using the fuzzy logical operator* **and**.

The **outer and product** of matrices can be computed as follows. Let the fuzzy set $A$ be represented by a column vector where each element is equal to the defined value of the membership function. Let the fuzzy set $B$ be represented in a similar way but as a row vector. Then their product is

$$\begin{bmatrix} a_1 \\ a_2 \\ \vdots \\ a_n \end{bmatrix} \times \begin{bmatrix} b_1 & b_2 & \dots & b_m \end{bmatrix} = \begin{bmatrix} a_1 \wedge b_1 & a_1 \wedge b_2 & \dots & a_1 \wedge b_m \\ a_2 \wedge b_1 & a_2 \wedge b_2 & \dots & a_2 \wedge b_m \\ \vdots & & & \vdots \\ a_n \wedge b_1 & a_n \wedge b_2 & \dots & a_n \wedge b_m \end{bmatrix} \tag{9.12}$$

---

**EXAMPLE 9.8** Example 9.7 continued

In this example let matrices $A$ and $B$ be equal to the fuzzy set of large positive error signal (**lpe**) and positive control signal (**pcs**), respectively.

$$\mathbf{lpe} = \begin{bmatrix} 0 \\ 0 \\ 0 \\ 0.6 \\ 1 \end{bmatrix} \qquad \mathbf{pcs} = \begin{bmatrix} 0 & 0.2 & 1 \end{bmatrix} \tag{9.13}$$

Again, recall that **and** is computed using the minimum of the degrees. Then the **outer and product** of these two vectors is as follows.

$$\begin{bmatrix} 0 \\ 0 \\ 0 \\ 0.6 \\ 1 \end{bmatrix} \times \begin{bmatrix} 0 & 0.2 & 1 \end{bmatrix} = \begin{bmatrix} 0 \wedge 0 & 0 \wedge 0.2 & 0 \wedge 1 \\ 0 \wedge 0 & 0 \wedge 0.2 & 0 \wedge 1 \\ 0 \wedge 0 & 0 \wedge 0.2 & 0 \wedge 1 \\ 0.6 \wedge 0 & 0.6 \wedge 0.2 & 0.6 \wedge 1 \\ 1 \wedge 0 & 1 \wedge 0.2 & 1 \wedge 1 \end{bmatrix} = \begin{bmatrix} 0 & 0 & 0 \\ 0 & 0 & 0 \\ 0 & 0 & 0 \\ 0 & 0.2 & 0.6 \\ 0 & 0.2 & 1 \end{bmatrix} \tag{9.14}$$

---

The **outer and product** is also known as **outer min product**. This name refers to the characteristic of logical operator on fuzzy sets. This operation has a great role in fuzzy control because it can be found in rules of most controllers.

### 2.3.3 INFERENCE ON FUZZY SETS

The rule-base of a (fuzzy) controller contains several rules in the form of implications $A \longrightarrow B$. If statement $A$ becomes true then we have to find all the rules containing this statement in their conditional parts. Collecting all these rules we have to conclude the necessary action(s). This method is called inferencing because we infer i.e. conclude facts from other facts. There is a frequently used inference method in Boolean logic, the modus ponens, which can be generalized to the case of fuzzy sets to obtain the *generalized modus ponens* (see section 1.2 of Chapter 3). The general form of the generalized modus ponens is as follows.

$$\frac{A', \quad A \longrightarrow B}{B'} \tag{9.15}$$

This means that if there is a rule $A \longrightarrow B$ in the rule-base and a fact $A'$ which is 'similar' to $A$ becomes true, the conclusion fact $B'$, which is almost the same as $B$, will also be true.

In the case of fuzzy controllers the statements in the conditional part are fuzzy sets and the similarity originates from the application of linguistic modifiers. The rules in modus ponens refer to relations between two fuzzy sets. So by applying the generalized modus ponens we can infer based on a relation and a fuzzy set to an another fuzzy set as it can be seen in the following definition.

**Definition 9.3. Compositional rule of inference**

*Let $R$ be a relation between universes $U_1$ and $U_2$ and $A$ a fuzzy set defined on $U_1$. Then the compositional rule is*

$$A \circ R = B \tag{9.16}$$

*where the resulting set $B$ is a fuzzy set on universe $U_2$ and $\circ$ is the composition operator.*

The composition operator is the inner matrix product defined in (9.7). The use of this rule is illustrated in the following example.

EXAMPLE 9.9 Compositional rule

Let relation $R$ be defined between the fuzzy sets **lpe** (large positive error) and **pcs** (positive control signal) of Example 9.7. Then this relation is an implication between these sets

$$\mathbf{lpe} \longrightarrow \mathbf{pcs} \tag{9.17}$$

and the result in matrix form is

$$R = \begin{bmatrix} 0 & 0 & 0 \\ 0 & 0 & 0 \\ 0 & 0 & 0 \\ 0 & 0.2 & 0.6 \\ 0 & 0.2 & 1 \end{bmatrix} \tag{9.18}$$

Let us apply the linguistic variable **somewhat** on the fuzzy set **lpe**:

$$\mathbf{somewhat\ lpe} = \begin{bmatrix} 0 & 0 & 0 & 0.6 & 1 \end{bmatrix}^{\frac{1}{2}} = \begin{bmatrix} 0 & 0 & 0 & 0.77 & 1 \end{bmatrix} \tag{9.19}$$

If we have a measurement record from the system which describes the degree of the error as a somewhat large positive value ($\mathbf{lpe}'$) then the necessary interaction $\mathbf{pcs}'$ can be calculated based on the relation $R\ :\ \mathbf{lpe} \longrightarrow \mathbf{pcs}$ of the rule-base as follows.

$$\begin{bmatrix} 0 & 0 & 0 & 0.77 & 1 \end{bmatrix} \circ \begin{bmatrix} 0 & 0 & 0 \\ 0 & 0 & 0 \\ 0 & 0 & 0 \\ 0 & 0.2 & 0.6 \\ 0 & 0.2 & 1 \end{bmatrix} = \begin{bmatrix} 0 & 0.2 & 1 \end{bmatrix} \tag{9.20}$$

## 3. RULE-BASED FUZZY CONTROLLERS

The overall structure of a rule-based fuzzy control system is shown in Fig. 9.13 [81], [86], [87], [88]. One can see that a fuzzy rule-based controller is a composite system. The controller consists of a preprocessing unit, a rule-base, a defuzzifier and a postprocessing unit. The task of preprocessing is to convert the error signal which is crisp data into a fuzzy form by calculating the difference between the reference input and system output. The next element, the rule-base is used for inferencing, i.e. for the determination of the necessary control action. The defuzzifier unit converts the determined fuzzy control action back into crisp value. As a last step the tuning and amplifying of the signal can be done by the postprocessing unit.

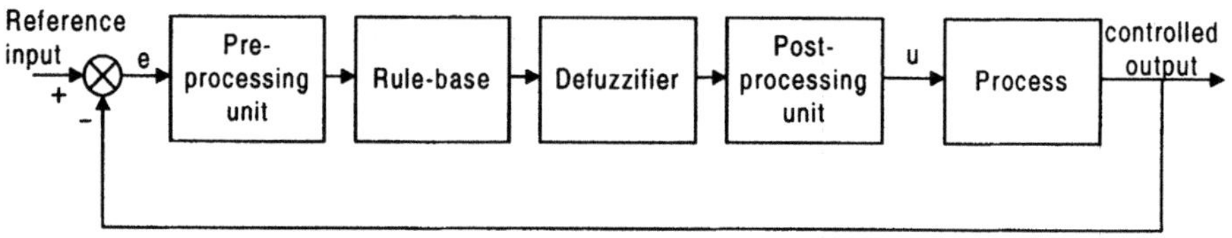


*Figure 9.13.* A fuzzy controller

Although this does not show from the figure, fuzzy controllers are very convenient tools for multi input - multi output process control, too.

This section describes the design steps and elements of fuzzy controllers.

## 3.1 DESIGN OF FUZZY CONTROLLERS

There are two main methods for the design of fuzzy controllers:

- *Direct controller design*: we design the fuzzy controller directly without modelling the process to be controlled.

- *Design of a process model*: we model the process to be controlled in a fuzzy way and use this fuzzy model to design the controller.

The two methods have similar steps, the difference is in the result of the modelling process: in the first case we get the fuzzy model of the controller while in the second case the model of the process.

There are different types of controllers developed for fuzzy control. The most important ones are the fuzzy PID controller, the table based controller, the self-organizing controller and the neuro-fuzzy controller. In the following we summarize the main steps of the design and the general characteristics of the elements of fuzzy controllers.

### 3.1.1 THE INPUT AND OUTPUT SIGNALS OF A FUZZY CONTROLLER

The selection of input and output signals of a fuzzy controller is a very important task because it has a great impact on the way universes, membership functions and rules are determined, i.e. it defines the structure of the controller. Typical inputs are the difference between the reference signals and the outputs of the controlled system, i.e. the error signals and the derivatives and integrals of the errors.

For proper selection we need some information about the nature of the system to be controlled. This information is related to system dynamics, stability, nonlinearity, time dependency of system parameters, etc. The type of controller can be selected on the basis of these data and the control goal.

As it was mentioned earlier, it is very easy to implement a fuzzy controller for MIMO systems. This fact enables us not only take the error signal and its changes into account, but also other signals, e.g. state variables and noises. Note that the increasing number of variables causes the rule-base to rapidly grow more complex. This is why it is useful to keep the number of variables on a reasonable level or to decompose the controller into subcontrollers, which are connected to each other either in a parallel or in a hierarchical manner.

The controlled input signal of the system can either be the absolute value or the incremental value of the control signal, similarly to crisp digital controllers. In the first case, the new position of the controller device is the result of the inference on the rule-base, while in the latter case the result is a change to the previous value.

### 3.1.2 THE SELECTION OF UNIVERSES AND MEMBERSHIP FUNCTIONS

As the next step in designing a fuzzy controller we have to determine the universes and membership functions for each variable.

The choice of universes depends on the system to be modelled. We have to determine the possible minimum and maximum values of the input signals of the fuzzy controller, i.e. the operating ranges of the measured output variables of the system. The selection of this range and its resolution has an impact both on the accuracy and on the calculation requirements.

The universes can be standardized for all variables. The usual standard ranges are the intervals $[-1, 1]$ where the real numbers of this interval are used and $[-100, 100]$ where the percentage of the actual value is referred to. For this we have to determine a scaling factor and a zero level for each signal to fit it to the selected range of the universe.

Having determined the universes we have to make a decision relating to the number and shape of the membership functions. The problem is similar to the selection of variables: if we use many membership functions for each variable then we need an exponentially growing number of rules in the rule-base. On the other hand, a small number of membership functions decreases the flexibility of the controller, especially in the case of nonlinear systems. The rule of thumb is to select three membership functions or in special cases two or five functions. In the case of three membership functions, the linguistic variables **small**, **medium** and **large** are used in general, while in the case of five functions modifier **very** is added to have **very small** and **very large**, too. If the universe is symmetric to the zero value then the linguistic variables **negative**, **zero** and **positive** (and **large negative/positive**) are used in general.

The other question is whether to use continuous or discrete membership functions. There are several shapes for continuous membership functions as it was mentioned in section 2. of this chapter. Continuous membership functions describe the changes of variables better but more time is needed for inferencing. The discrete membership functions are given as vectors. Inferencing is easier in this case but the number of vector elements influences the accuracy.

If we have any a priori knowledge about the shape of membership functions we can use it. In other cases we can select from the ones mentioned ones in section 2. of this chapter.

Nowadays, a scalar rather than a fuzzy set is used frequently as an input value of a fuzzy controller, which is an output signal of the system or an error signal being the difference of the reference value and the output signal. The scalar controller input is called a singleton and it can be considered as a special fuzzy set where the grade of membership can either be equal to 1 or to 0. The main advantages of application of singletons are as follows:

- inferencing is simpler;

- it makes the writing of rules more intuitive.

To summarize the selection of membership functions we recommend the use of the following steps as a rule of thumb:

- Let the number of membership functions be 3. As first approximation three sets are enough to cover the lower, medium and upper zones of the variables. Later on we can add more sets based on operational experiences.

- Select a triangular shape for each membership function. These triangles should be symmetrical and similar for each variable. The leftmost and the rightmost should be shouldered ramps (see Fig. 9.4).

- The base of these triangles should be so wide that it allows each value of the universe to be a member of two sets at least. If there is a gap between two sets then there is no rule for the values in the gap. If a given value is a member of more than one set then the application of more rules makes control smoother and more flexible.

### 3.1.3 THE RULE-BASE

The rule-base contains the rules for operating fuzzy controllers.

The most important task is to find the suitable rules for the controller. In general we can select from the following possibilities to find the rules (they can also be combined if necessary):

- *Using a normalized or standard rule-base*
  In this case the error signal and its derived and/or integrated values are used as a fuzzy PID (or P, PD, PI) controller. When scaling the input and output values to a given universe we can use tables like this below in the case of a PD controller to compute the control signal (the controlled input of the system):

| | | $\Delta e$ | | | | |
|---|---|---|---|---|---|---|
| | | **LN** | **SN** | **ZERO** | **SP** | **LP** |
| | **LN** | ln | ln | sn | sn | nc |
| | **SN** | ln | sn | sn | nc | sp |
| $e$ | **ZERO** | sn | sn | nc | sp | sp |
| | **SP** | sn | nc | sp | sp | lp |
| | **LP** | nc | sp | sp | lp | lp |

  where **ln** refers to *large negative*, **sn** refers to *small negative*, **nc** refers to *no change*, **sp** refers to *small positive* and **lp** refers to *large positive* manipulated variable value.

  Each element of this table is a rule. For example the third row and the second column refers to the following rule:

  ```
  If the error signal is equal to zero
  and the change in the error signal is small negative
  then the control signal is small negative.
  ```

  Note that the main advantage of a fuzzy controller is not its ability to simulate a linear controller but the easy and understandable way it controls nonlinear systems. At the same time, fuzzy controllers make the dynamic behaviour of controlled linear systems smoother because they are not too sensitive to noise. If we know the parameters of a linear controller we can use them as initial parameters for a fuzzy controller thus making the tuning of the fuzzy controller simpler.

- *Using the experience and intuition of experts*
  Rules can be derived from the operator's handbooks and logbooks

of the plant. They can also be set up as a result of interviewing the operators. The latter can be done by using a carefully designed questionnaire to collect the rules of thumb related to the system to be controlled. It is also very useful to observe an operator's control actions and deduce **if-then** type rules.

- *Using the fuzzy model of the process*
  As it was mentioned the fuzzy model of the process can be used to obtain the rule-base of the controller. The model of the system can be viewed as a special inverse of the model of the controller.

- *Using learning type controllers*
  Some special fuzzy controllers like self-organizing and neuro-fuzzy controllers can amplify and correct their own rule-base.

Although a rule-base contains the rules in an **if-then** format they can be presented to the end-users in different ways. Besides the linguistic description, relational or tabular format and graphic representation are also frequently used.

### 3.1.4 THE RULE-BASE ANALYSIS

As we could see from the previous sections, the rule-base plays a central role in fuzzy control. A well designed rule-base is the main requirement of the proper operation of fuzzy control.

In this section the following properties are investigated in connection with the fuzzy rule-base [89]:

- completeness,
- consistency,
- redundancy,
- interaction.

**Completeness.** A rule-base is *complete* if every non-zero input generates a non-zero output. In the case of fuzzy sets the non-zero input/output refers to a fuzzy set with only zeros as elements.

There are two main reasons for the incompleteness of a rule-base. In the first case, there is a gap between membership functions. This is easy to check with the help of the graphic representation of membership functions. In the second case, one or more rules are missing. It is much more difficult to discover this, especially in the case of large, complex rule-bases.

One of the simplest and quickest methods of *checking the completeness of a fuzzy rule-base* is as follows.

Assume that there is no indefinite fuzzy set for the output signals of the system to be controlled, i.e. every value of the universe of the output signal belongs to at least one membership function. The graphic representation of the membership functions will show this. If this assumption holds then it is enough to check the conditional parts of the rules. Assuming that the controller has $n$ inputs (which are the system outputs) then the input space of a fuzzy controller denoted by $X$ is a cartesian product of all the possible input values. Let us denote the conditional part of the $i$-th rule by $X_i$, a fuzzy set in $X$ by $\mu_x$, the inference part of the $i$-th rule by $U_i$ and the number of rules by $n$. Then the general form of a rule in the rule-base is:

$$\textbf{if } X_i \textbf{ then } U_i, \qquad \text{where } i = 1, \dots, n.$$

The controller is complete if

$$\forall x \in X : \exists \; X_i(x) > \varepsilon \qquad \text{where } 1 \leq i \leq n \text{ and } 0 < \varepsilon \leq 1 \tag{9.21}$$

According to this relation a rule-base is complete if there exists at least one rule which contributes to the output by a number larger than $\varepsilon$.

If the variables of the conditional parts of the rules are combined using only the operator **and** then the completeness can be tested by checking the validity of the inequality

$$(\bigvee X_i) > \varepsilon \tag{9.22}$$

**Consistency.** A rule-base is *inconsistent* if two or more rules with the same or very similar conditional parts generate different outputs. These different outputs cause more than one peaks in the curve which is the graphic representation of the fuzzy set given by the inference engine of the controller.

In the case of a consistent rule-base all the rules with slightly different input parts have to generate slightly different output sets. This means that there is a need to measure the differences between input and output parts. The next comparison is introduced in literature:

$$m_{ij} = (X_i \textbf{ similar\_to } X_j) \textbf{ and } \textbf{not}(U_i \textbf{ similar\_to } U_j) \tag{9.23}$$

where the operation **similar_to** computes the degree of similarity between two fuzzy sets. One of the easiest methods to decide on similarity is to compute the overlap between the two fuzzy sets in a **similar_to** relation.

The result of consistency checking is a symmetric matrix $M$ with a size $n \times n$ and the $ij$-th entry $m_{ij}$ refers to the inconsistency between rules $i$ and $j$. The larger the value $m_{ij}$, the larger the inconsistency.

**Redundancy.** A rule is redundant if there is at least one other rule in the rule-base with the same very similar **if-then** parts. There can be two reasons why a rule-base contains redundant rules. The simpler case is when the user, by mistake, adds the same rule twice to the rule-base. The other source of redundancy is a new rule to be added to the rule-base, but already covered by an existing rule.

Although the redundancy itself does not cause inconsistency, it can lead to it thus causing a growing demand on storage and computing time.

To check redundancy, the sets of rules have to be compared. A rule is redundant if its sets are subsets of another rule. This can be expressed as follows:

$$\mu_i = R_i \textbf{ in } (R \setminus R_i) \tag{9.24}$$

where $R_i$ is a rule in rule-base $R$ (where $R = \bigvee R_j, \quad j = 1 \ldots n$). To measure redundancy the way we determine $R$ is modified as follows:

$$R' = R \setminus R_i = \bigvee R_j, \quad j = 1 \ldots n \text{ but } j \neq i \tag{9.25}$$

In order to compare between rule $R_i$ with the other part of the rule-base rule $R_i$ is transformed into a matrix, which is the outer product of its input and output parts.

The operation **in** can be done easily by comparing matrices. If the elements of matrix $R'$ in Eq. (9.25) are greater or equal to the elements of matrix $R_i$ then rule $R_i$ is redundant.

**Interaction.** Interaction is related to the independency of the conditional parts of rules. If the input relations of these conditional parts are disjoint then there is no interaction between the rules in the rule-base.

The overlap between the input relations can cause interaction in the following way. Although an input instance is exactly the same as the conditional part of a rule, $R_i$, the inferred output set may not be equal to the output part of this rule. The reason of this difference is the interaction between rule $R_i$ and other rules in the rule-base, that is, the input relation can be matched to more than one conditional part of rules and so the inferred fuzzy set is a combination of the output parts of these rules.

Having no overlap between the input sets does not belong to the general requirements but it can be useful to measure the degree of interaction.

The degree of interaction can be measured by

$$\nu_i = \|(X_i \circ R) - U_i\| \tag{9.26}$$

where $X_i$ and $U_i$ are the input and output parts of a rule $R_i$, respectively, $R$ refers to the rule-base ($R = \bigvee R_i$), $\|.\|$ is a suitable vector norm (or fuzzy set norm) and $\nu_i$ is the degree of interaction between rule $R_i$ and the rule-base $R$. The larger the value of $\nu_i$ is, the more interaction there is between them.

## 3.2 THE OPERATION OF FUZZY CONTROLLERS

In the previous three sections we described the basic components of a fuzzy controller. With these elements, we can start operating it. Here the main units of fuzzy controllers are described in more details.

### 3.2.1 THE PREPROCESSING UNIT

The main task of a preprocessing unit is to convert the output signals coming from the system into input data for the inferencing process in the rule-base. These input data are the grades of membership for the conditional parts of the rules. To carry out the conversion the values of the input signals of the controller (that is the output signals of the system) have to first be scaled to the standardized universes. Then grades of membership have to be determined for all membership functions related to the given variable.

This process is often referred as fuzzification.

### 3.2.2 THE INFERENCE ENGINE

Using the fuzzy inference we can determine to what extent each rule is fulfilled. If the conditional part of a rule contains more than one condition (in **and** relation) then the function **min** is used to compute the grade of the conditional part as it was shown in section 2.2 of this chapter.

Inferencing consists of the following steps (illustrated in Fig. 9.14).

Assume the following rules:

**If** $e$ `is small negative` **and** $\Delta e$ `is large negative`
**then** $u$ `is large negative`

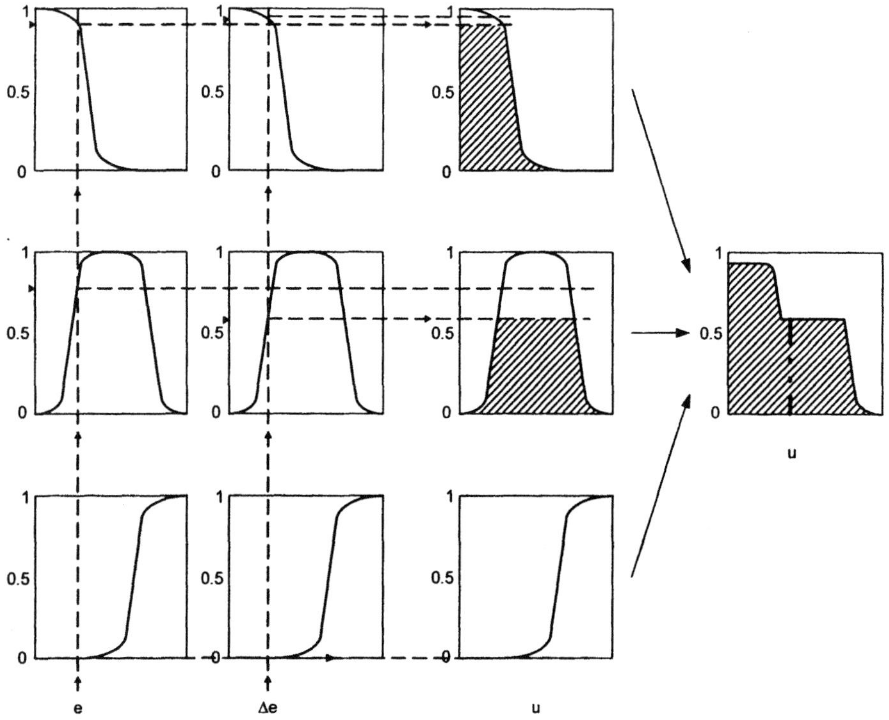


*Figure 9.14.* The fuzzy inference procedure

**If** $e$ `is zero` **and** $\Delta e$ `is large negative`
**then** $u$ `is small negative`

These rules can be derived from the table defined in section 3.1.3 of this chapter but for the sake of simplicity we assume that the other rules there have no contribution to the final value of the control signal, that is, the manipulated input variable of the system.

**Step 1** is done in the preprocessing unit when the membership grade is determined. This is illustrated by vertical lines in the first and second columns on the left in Fig. 9.14.

**Step 2** The inference engine determines the membership grade of each term in the conditional parts of the rules. This is shown by horizontal lines in the first and second columns in Fig. 9.14.

**Step 3** Using the operation **min** (fuzzy **and**) the inference engine determines the grade of fulfillment for the conditional parts of each rule and implies the contribution of the rule to the output value. This is depicted by the shadowing in the third column.

**Step 4** Collecting all contributions and using operation **max** (fuzzy **or**) the resulting fuzzy set is determined which is shown in the fourth column of Fig. 9.14.

**Step 5** The resulting fuzzy set has to be converted into a crisp value for the controlling element. There are several methods to do this, some of them are described in section 3.2.3 of this chapter below. Using the centre of area method the crisp value is shown in the graph of the fourth column.

In Steps 3 and 4 we used the **max** – **min** operation introduced in section 2.2 of this Chapter. However, there are other implication methods in literature. *Star-implication* uses multiplication rather than the operation **and**. It results in a slightly smoother control signal because multiplication more or less preserves the original shape of membership curves.

For singleton type outputs, *sum-star inference* is used. Its result is equal to the linear combination of singletons and their contribution to the output value derived from the rules in Step 3.

### 3.2.3 THE POSTPROCESSING UNIT

The main task of the postprocessing unit is to convert the fuzzy set given by the inference engine into a crisp control signal. This process is called defuzzification.

The most important methods are as follows.

1. *Mean of maxima*

   This method determines the crisp control value as the maximum possible value, i.e. the maximum grade of membership. If there are more than one maximum points then it calculates their average as follows.

$$u = \frac{\sum_{j=1}^{l} x_{mj}}{l} \tag{9.27}$$

   where $x_{mj}$ denotes the maximum value of the $j$-th term in the resulting fuzzy set, and $l$ is the number of terms.

2. *Centre of area method*

   In this case the defuzzification process calculates the value which divides the resulting fuzzy set into two parts with equal areas. In the case of discrete membership functions this point can be calculated on the basis of the following formula.

$$u = \frac{\sum_{j=1}^{l} \mu(x_j) \cdot x_j}{\sum_{j=1}^{l} \mu(x_j)} \tag{9.28}$$

where $\mu(x_j)$ is the membership grade of the $j$-th term at the value $x_j$ of the discrete universe.

3. *Selecting the maximum value*
   One of the simplest defuzzification methods is to select the term with the maximum membership grade. The variations of this method select the leftmost maximum (called *first of maxima* or *FOM*) or the rightmost maximum (*last of maxima* or *LOM*).

4. *Height*
   For singleton type outputs the steps of inference and defuzzification can be combined as follows.

$$u = \frac{\sum_{j=1}^{l} \alpha_j \cdot s_j}{\sum_{j=1}^{l} \alpha_j} \tag{9.29}$$

where $s_i$ is the value of the $j$-th singleton and $\alpha_j$ is its weight in the given rule.

Chapter 10

# G2: AN EXAMPLE OF A REAL-TIME EXPERT SYSTEM

G2 of Gensym [90], [91] is an excellent graphical, object-oriented environment for rapid prototyping and implementing real-time expert systems. At the same time it exhibits almost all features and properties of a real-time expert system shell in a very transparent and user-friendly way.

The general notions and concepts, as well as the background material about real-time expert systems is given in Chapter 6.

The following characteristics of G2 are described in this chapter.

- Knowledge representation in G2
- The organization of the knowledge base
- Reasoning and simulation in G2
- Tools for developing and debugging knowledge bases

It is important to emphasize that the material in this Chapter is by no means a comprehensive and extensive introduction into G2, neither is its User Manual. The aim here is to illustrate the most important concepts, tools and techniques on an excellent example of a real-time expert system. The interested Reader is referred to the manuals of G2 for all details and for a comprehensive description.

The components of G2, together with the development and operation of a knowledge base are illustrated with the example of the batch water heater system (coffee machine) introduced in Appendix B.

# 1. KNOWLEDGE REPRESENTATION IN G2

The application development in G2 is assisted by a well-structured natural language in a high-level, intuitive and graphic-oriented development environment. This environment promotes rapid prototyping with the help of predefined knowledge base elements and refining to an adequate full-sized real-time system.

The initial step in G2 adaptation is to define the *class* of each *object* that appears in the application: what it looks like, what its typical attributes are and how they can be connected to other objects. Thereafter a concrete model is planned by placing *objects* in one (or more) *workspace(s)* and connecting them to show their relationships. The result is a schematic diagram of the application like the one in Fig. 10.1 of the coffee machine (bath water heater system).

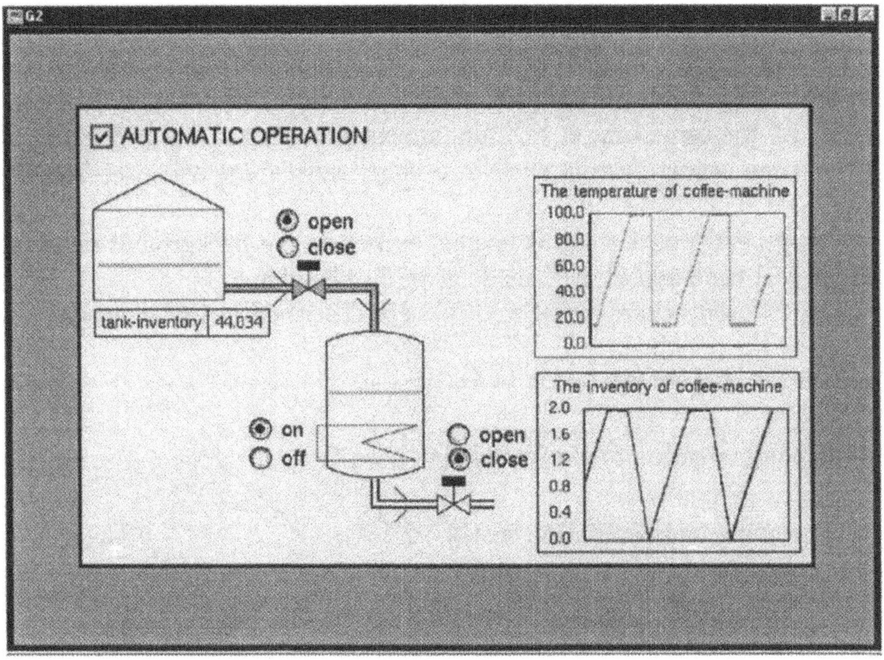


*Figure 10.1.* **The schematic diagram of coffee machine system**

Every object in the schematic diagram has a *table* with its properties. These attribute tables are automatically generated by G2 from the definition of the class of the object.

There are two specific object types that represent changeable data: ***variables*** and ***parameters***. A variable has a validity interval associated with it. Whenever G2 needs the value of a variable after its validity has expired, it automatically gets it from the ***data source*** or ***data server*** of the variable. This data server may be the G2 inference engine, the G2 simulator or an external data source like a sensor, an external database or a user. A parameter differs from a variable in that it must always have value. This means a parameter needs to have an initial value. Its value can be changed by rules, formulas or procedures.

***Rules*** represent the expert's knowledge. They describe how to reason and respond to a given set of conditions. They are used to conclude the value of some variables by the real-time inference engine, to show how G2 responds and what it concludes from changing conditions within the application. They can be event-driven (through forward chaining) to automatically respond whenever a new data item arrives, and can be data-driven (through backward chaining) to automatically invoke other rules, procedures or formulas. A natural language context-sensitive editor is used for entering the rules and other text. It is good to make rules as generic as possible in order to use them as little as possible.

A complex sequence of actions can be performed in a cycle by accident until certain conditions come true. Such sequences are best represented by G2 ***procedures***. Like rules, procedures may ask G2 to execute some task and unlike rules, they do not response to conditions but define an instruction sequence. They resemble to procedures found in several structured programming languages.

Some variables and parameters can receive values from the ***G2 simulator***. In this case the developer needs to create simulation formulas that tell G2 how to find the simulated values. These formulas can be algebraic, difference and first-order differential equations. Simulation formulas are used for defining complex, high-order models and these models may be either linear or non-linear. The G2 simulator can be used for modeling and simulating data that cannot be measured. It is possible to compare data from an external data source with the simulated values in order to diagnose the failure of an operation and to test the application while it's being developed.

While some objects and connections are permanent in an application, there may be transient objects and connections, too. These are generated and deleted by certain actions which are contained, for example, by rules and procedures. The transient objects and connections aren't saved in the knowledge base.

The end-user needs to get a lot of different information and needs to respond to them during the run-time of an application. G2 has several predefined objects that help communication: ***end-user controls*** like check-boxes and buttons; ***displays*** like graphs and meters,which show the values of variables, parameters or expressions; a ***logbook*** that informs the user about system conditions, errors and warnings; and a ***message board*** that shows the messages of G2.

The knowledge base can be separated into any number of ***workspaces*** by the developer. For example, there can be a workspace for rules, another for class definitions, another for the schematic diagram and so on. Any object and object definition may have its subworkspace. A subworkspace can hold items that in turn have their own subworkspaces, and so on. In this way knowledge can be organized hierarchically.

The *items* created by the developer as object classes, objects, rules, procedures, formulas, workspaces etc. make up the ***knowledge base*** for the application. In most applications the knowledge base is built up gradually. The first step is to develop and test a prototype within a few hours. The full-sized application then evolves from refining and refining the prototype.

After the knowledge base is built, it can be connected with *external data sources* using the data interfaces available for G2.

## 2. THE ORGANIZATION OF THE KNOWLEDGE BASE

A knowledge base contains knowledge about a given application in the form of the following special components:

- objects: aims of interest in an application
- object definitions: definitions of object classes that appear in the knowledge base
- workspaces: contain the objects, connections, rules etc. in an application
- variables and parameters: special objects that represent changing values
- connections and relations: physical, logical and other relationships among objects
- rules: knowledge of how to reason and respond to a given set of conditions

- procedures: instruction sequences
- functions: built-in or user-defined operations

## 2.1 OBJECTS AND OBJECT DEFINITIONS

An *object* is a representation of a part of an application, in the case of the coffee machine, the water-tank and the valves in the physical world are represented in G2 by objects named `vessel`, `atmospheric-tank` and `valve`. Fig. 10.1 shows the schematic representation of the objects connected in the coffee machine. These objects are generated manually by the developer and they exist permanently in the knowledge base. The transient objects generated by rules or procedures only exist when the knowledge base is running.

The picture that graphically represents an object is called an *icon*. The pipes and wires that connect objects are called *connections*. As Fig. 10.2 shows, each object has an attribute table with two columns. The first contains the attribute names and the second the attribute values or stars when the variable has no value. For example, the attribute table of a `vessel` contains knowledge about its `names`, `inventory`, `capacity`, and so on. Attributes defined by any type of variable or parameter have sub-tables that describe their properties.

Every object belongs to a class and classes exist within a hierarchy. Each class in the hierarchy inherits the attributes, icons and connection stubs of its superior class, but it may also have its own class-specific attributes, its own unique icon and connection stubs. For example, a `coffee-machine` belongs to the `vessel` class. As it can be seen in Fig. 10.3, the direct superior class of `vessel` in the object-definition table is the `container-or-vessel` class, which belongs to the `process-equipment` class, which in turn belongs to the `object` class, which in turn belongs to the `item` class. A `vessel` has four inherited attributes, has no class specific attribute, but has its own icon and stubs.

The object classes used in the coffee machine system and its class hierarchy appear in Fig. 10.4. `Valve-1` and `valve-2` both are instances of the `valve` class. Objects in the same class have the same icons and attributes, but of course attribute values may be different.

The class hierarchy is part of the item hierarchy, where the items (objects, workspaces, rules, procedures, etc.) are organized into classes. The item hierarchy determines how G2 applies its generic expressions. For example, a generic rule that begins with `for any object` applies to all objects and all subclasses of the main object class in the knowledge base.

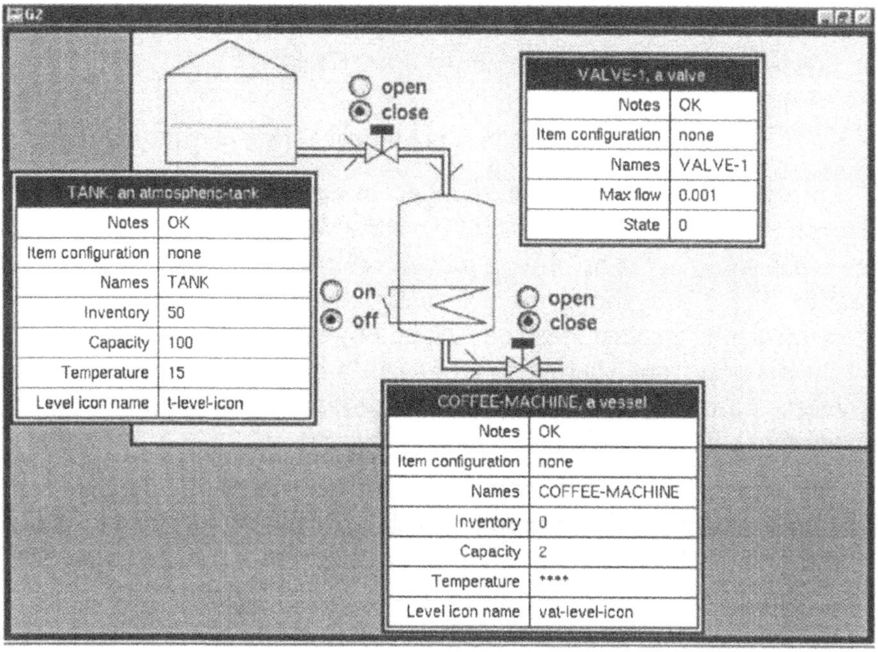


*Figure 10.2.* **Attribute tables**

## 2.2 WORKSPACES

***Workspaces*** are rectangular areas that contain all types of items (objects, connections, rules, and so on) except workspaces in an application. The knowledge base elements are placed in any number of workspaces, which may be top-level workspaces and subworkspaces. A subworkspace is a workspace that is associated with an object, object definition or connection definition. It may have some subworkspaces of its own, too. This hierarchy of workspaces makes it possible to organize the knowledge hierarchically. In addition, it is possible to activate and deactivate a workspace (and all of its items) selectively. The rules, objects and any items of a deactivated workspace are ignored by the inference engine until the workspace is reactivated again.

Besides permanent workspaces there are temporary workspaces, which are not elements of the knowledge base. They only exist when the knowledge base is running and are not saved with it.

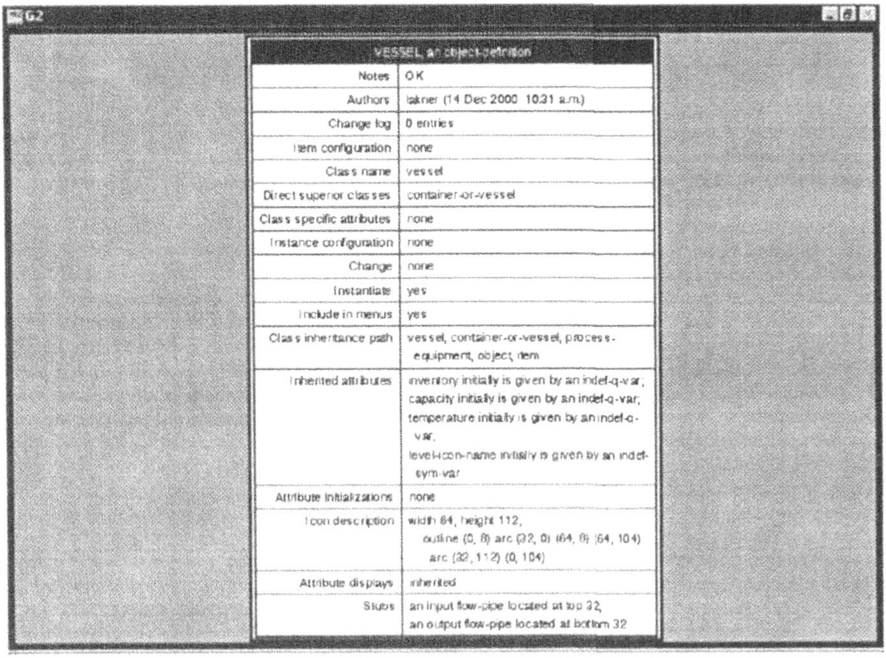

VESSEL, an object-definition

| | |
|---|---|
| Notes | OK |
| Authors | lakner (14 Dec 2000 10:31 a.m.) |
| Change log | 0 entries |
| Item configuration | none |
| Class name | vessel |
| Direct superior classes | container-or-vessel |
| Class specific attributes | none |
| Instance configuration | none |
| Change | none |
| Instantiate | yes |
| Include in menus | yes |
| Class inheritance path | vessel, container-or-vessel, process-equipment, object, item |
| Inherited attributes | inventory initially is given by an indef-q-var; capacity initially is given by an indef-q-var; temperature initially is given by an indef-q-var; level-icon-name initially is given by an indef-sym-var |
| Attribute initializations | none |
| Icon description | width 64, height 112, outline (0, 8) arc (32, 0) (64, 8) (64, 104) arc (32, 112) (0, 104) |
| Attribute displays | inherited |
| Stubs | an input flow-pipe located at top 32, an output flow-pipe located at bottom 32 |

*Figure 10.3.* **Object definition table**

## 2.3 VARIABLES AND PARAMETERS

*Variables* and *parameters* are used for representing values that change in time. In the coffee machine system for example,the **`temperature`** and the **`inventory`** of the **`coffee-machine`** are described with variables and the **`states`** of **`valves`** are described with parameters. This two special object types are similar in several points of view: they may have attributes, they may be organized into classes and icons may belong to them. In addition, both of them have a **`history keeping spec`** attribute, which tells G2 whether to keep or not to keep a history of values. Having compiled a history of values, G2 is able to provide information on stored data, e.g. average and maximum values, rate of changes etc.

The main difference is that while a parameter always has to have a value, the value of a variable may expire. The **`validity interval`** attribute of the variable defines an interval over which the last recorded value is valid. As G2 needs to find new values for variables, every variable has a data source or data server which automatically rereads it. The data seeking techniques may be:

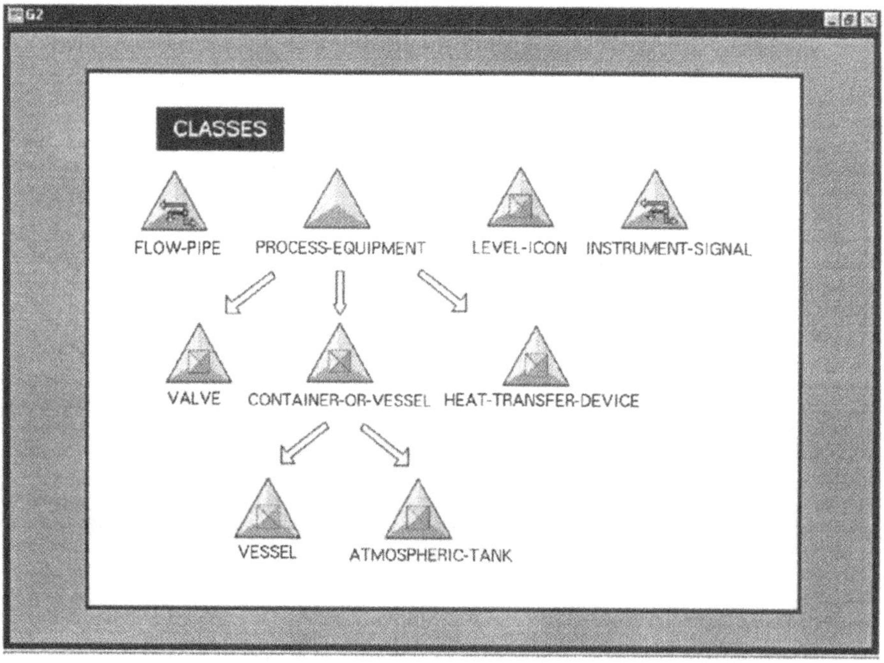


*Figure 10.4.* **Class hierarchy**

- reading the value from an ***external data source***
- receiving the value from a ***G2 simulator***
- inferring the value from the rules in the ***G2 inference engine*** using backward chaining

Variables can also have specific formulas and simulation formulas which G2 can use to calculate their values.

G2 never needs to search for a value of a parameter as it is guaranteed to always have a current value and unlike a variable, a parameter must have an initial value. Its value can be changed by rules, procedures, formulas or simulation formulas.

## 2.4 CONNECTIONS AND RELATIONS

The conjunctive pipes and electrical wires between objects in a schematic diagram are called *connections*. A connection is an item that graphically links two objects in order to indicate the relationship between them.

In G2, the developer can define a class of connections, he can graphically link objects to each other, he can refer to and infer objects and connections using their linking definitions. This makes it possible to write generic rules that refer to, for example, any `container-or-vessel connected to any valve`.

*Relations* are similar to connections in that they can be used to link objects. A relation is an association between two objects. The developer can define relation classes, can control the existence of a given relation between two objects and can conclude by existing relations.

The main differences between relations and connections can be summarized as follows:

- connections are constructed manually, but relations are defined dynamically
- relations do not have a graphical representation and they do not belong to the knowledge base
- while relations may exist between any type of units, connections only exist between objects

## 2.5 RULES

The expert's knowledge that describes how G2 should respond and answer to various conditions in an application is stored in *rules*. As described in section 2.2 of Chapter 2, a general rule in G2 has two parts: an *antecedent* or *condition* representing the conditions, and a *consequent* or *consequence* specifying what to do when the antecedent of the rule is true. The consequent of any rule contains actions, like `conclude`, `change`, `start`, and so on. Rules are invoked by G2's inference mechanism. The logical expression in the condition part is evaluated first. When one or more variables in the antecedent part do not have current values, G2 tries to get them from its data source or data server. If the antecedent part of the examined rule is true, G2 executes the actions in the consequent part.

From the operational point of view, rules can be grouped into five main categories in G2:

- `if` rules are common rules

```
for any valve V
if the state of V = 1
then change the center stripe-color of every flow-pipe
connected to V to sky-blue
```

- **when** rules are similar to if rules, except that, by default, G2 does not invoke a **when** rule through forward or backward chaining

```
for any container-or-vessel CV
when the value of the inventory of CV = 0
then conclude that the temperature of CV has no value
```

- **initial** rules are invoked only when the knowledge base starts or restarts

```
initially for any container-or-vessel CV
if the inventory of CV > 0
then conclude that the temperature of CV = 15
```

- **unconditional** rules are rules without antecedent part

```
initially for any valve V
unconditionally conclude that the state of V = 0
```

- **whenever** rules are driven only by events, for example when a variable or parameter receives a value

```
whenever auto-manual-state receives a value and
when the value of auto-manual-state is auto
then start auto()
```

The rules that contain the word any in the examples above are generic rules, which can be applied to more than one item in an application.

An attribute table of a rule is illustrated in Fig. 10.5. Some of the interesting attributes:

- `options` - available for rules to control how they are invoked
- `scan interval` - tells G2 how often to invoke the rule
- `focal objects` and `focal classes` - denote the specific objects and classes associated with the rule
- `rule priority` - used for scheduled rules
- `depth-first backward chaining precedence` - sets the order in which G2 looks at the rules in depth-first backward chaining
- `timeout for rule completion` - determines how long G2 may try to evaluate the antecedent of a rule

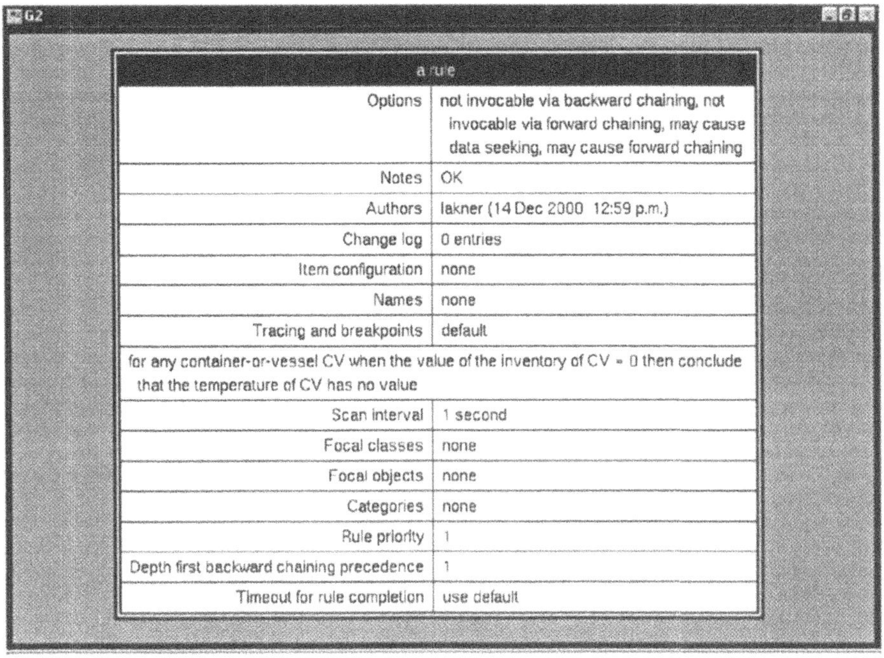

| a rule | |
|---|---|
| Options | not invocable via backward chaining, not invocable via forward chaining, may cause data seeking, may cause forward chaining |
| Notes | OK |
| Authors | lakner (14 Dec 2000 12:59 p.m.) |
| Change log | 0 entries |
| Item configuration | none |
| Names | none |
| Tracing and breakpoints | default |
| for any container-or-vessel CV when the value of the inventory of CV = 0 then conclude that the temperature of CV has no value | |
| Scan interval | 1 second |
| Focal classes | none |
| Focal objects | none |
| Categories | none |
| Rule priority | 1 |
| Depth first backward chaining precedence | 1 |
| Timeout for rule completion | use default |

*Figure 10.5.* **Attribute table of a rule**

## 2.6 PROCEDURES

A *procedure* is a series of operations or commands executed in sequence by G2. Procedures may be practically used in the following:

- sequential processing
- scheduled events
- complex control algorithms
- calculations containing actions
- same operations on different data values or on many occasions

A user-defined procedure in its attribute table is illustrated in Fig. 10.6. As it can be seen, the language of G2 procedures compares to the that of high-level programming languages. G2 contains all of the fundamental programming structures like conditions, iterations and it has several statements like `do in parallel` for real-time programming.

A procedure consists of three main parts:

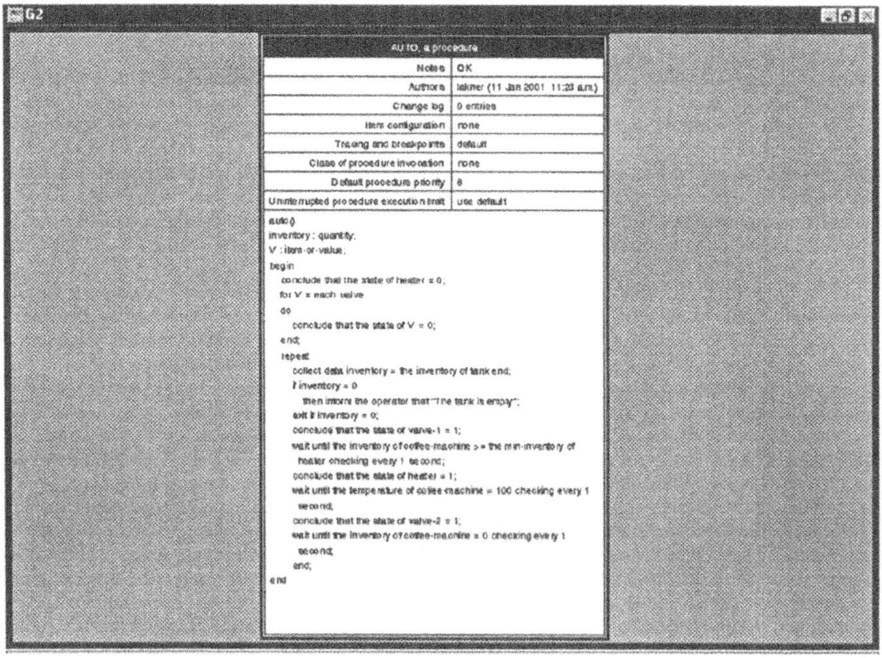


*Figure 10.6.* **Attribute table of a procedure**

- name, arguments and returns values (if any) of the procedure are defined in *procedure header*
- local variables with their types and initial values are specified in *local declarations*
- procedure statements are stored in *procedure body* nested in a begin-end block

## 2.7 FUNCTIONS

*Functions* are predefined, named sequences of operations. A function is called when its name and arguments (if any) appear as part of an expression and it returns a value. For example, the following are arithmetic function calls that return a number:

```
sqrt(x+y)
max(x,y,z)
abs(x)
```

G2 has several built-in functions and enables the construction of user-defined algebraic, logical and text functions, too. Besides these, it also has a foreign function interface, which is used for calling C and Fortran functions within G2.

# 3. REASONING AND SIMULATION IN G2

## 3.1 THE REAL-TIME INFERENCE ENGINE

The most powerful element of G2 is its inference mechanism. The *real-time inference engine* reasons the current state of the application, communicates with the end-user and initiates other activities based upon what it has inferred. It operates using the following sources of information:

- knowledge contained in the knowledge base
- simulated values
- values received from sensors and other external sources

The inference engine has the following abilities:

- *scanning* rules: it repeatedly invokes rules at regular time intervals, which are predefined by the `scan interval` attributes of the rules
- *focusing* on rules: a rule may be related to objects or classes by its `focal objects` or `focal classes` attribute, and executing a `focus` action on an object, G2 invokes all rules associated to it
- *invoking* rules: rules can be grouped into categories based on their `focal category` attributes, and G2 may invoke all rules in a category by the `invoke` action
- *wakeup* rules: when a variable that has been waiting for a value receives a value, the inference engine re-invokes the rule that was waiting for the value of the variable
- *data seeking*: when G2 needs the value of a parameter and this value has expired, G2 gets a new value from the appropriate data server, which may be the inference engine, the G2 simulator or other external data servers
- *backward chaining*: if the value of a variable is not given by any sensors or formulas, the inference engine uses backward chaining to infer it from rules (Section 3. of Chapter 3 discusses this chaining mechanism in detail)

- *forward chaining*: the inference engine uses forward chaining to invoke a rule when at least one of the conditions in its antecedent is satisfied by another rule (further information on forward chaining can be found in section 2. of Chapter 3)

Most inference engines have backward and forward chaining mechanisms, but the G2 inference engine has additional, essential techniques for working with real-time applications.

## 3.2 THE G2 SIMULATOR

The *G2 simulator* is a built-in part of G2, but it may be seen as an independent software unit or as a special kind of data server that provides simulated values for variables and parameters. It has the following most important properties.

- It is strongly connected with the other parts in G2. For example, the developer may define a specific simulation formula in the simulation subtable of a variable or may create a generic simulation formula as a statement of a workspace, like a rule.
- It is able to solve algebraic, difference and first order differential equations.
- It can assign individual simulation times to the different variables.
- Variables may have specific simulation formulas, but the classes of variables and parameters may have generic simulation formulas.
- It may run parallel with other real-time processes, so it can provide simulated values while G2 is controlling real operations.

The main aim of a G2 simulator is to test and provide simulated values: it can be used for testing the knowledge base during normal system operation or in the care of an obscure failure, it can simulate the occurrence of rare states while speeding up simulation time, it can estimate states that cannot be easily observed by sensors and it can simulate the operation of an application before on-line operation.

Three categories of variables can get values from the G2 simulator:

- *dependent* variables for algebraic equations:

  ```
  height * diameter * pi
  ```

- *discrete state* variables for difference equations:

  ```
  state variable:  next value = the inventory of tank -
  the max-flow of valve-1 * the state of valve-1,
  with initial value 100
  ```

- *continuous state* variables for differential equations:

```
state variable:  next value = - the max-flow of valve-1 *
the state of valve-1, with initial value 100
```

State variables depend on their previous values, so they must have initial values. On the other hand, dependent variables are functions of their actual values and simulated values of other variables. These variable categories are not explicitly defined, they are derived from the simulation formulas of the variables.

# 4. TOOLS FOR DEVELOPING AND DEBUGGING KNOWLEDGE BASES

## 4.1 THE DEVELOPERS' INTERFACE

An expert system is built up and run by the developer with the help of the *developers' interface*. The G2 developers' interface has the following main properties.

- It provides a graphic representation of the application, which is easily interpreted and used.
- It describes knowledge using a language very similar to English.
- It has a multiple text editor, which is used to enter and edit texts.
- It has an icon-editor to generate and modify icons of the objects.
- It has several tools for building, modifying and using large and complex knowledge bases.
- It can insert documentation into the knowledge base.
- It can help to release mistakes in rules, functions and formulas.

### 4.1.1 THE GRAPHIC REPRESENTATION

Building an application starts with generating its graphic model. Objects are represented with icons and unique icons may be defined for each object class. The developer models an application by locating and connecting objects on a workspace in a way that represents their relations. The result is a schematic diagram of the application.

When a knowledge base item (objects, connections, variables, rules, workspaces and so on) is clicked, a pop-up menu appears.It lists all the operations that developers and users can perform. Examples of operations are deleting, changing size and color, transferring, and so on.

Beyond it, every item has an attribute table which defines its properties. The attribute values can be defined and changed in the attribute table before the application starts running and even dynamically, during running.

### 4.1.2 G2 GRAMMAR

As we can see from the description of rules and procedures in sections 2.5 and 2.6 of this Chapter, G2 grammar is structured like the English language. It is important that this language can refer to items in several ways:

- by name:

  `coffee-machine`

- by class name:

  `the vessel`

- as the instance of a class that is nearest to another item on a schematic diagram

  `the level-icon nearest to coffee-machine`

- as the instance of a class that is connected or related to another item or class of items

  `the valve connected at the output of coffee-machine`

- a set of items is referred to using the **for** prefix, **any** and a class name:

  `for any valve`

  G2 grammar enables the use of generic rules and formulas:

  ```
  initially for any valve V
  unconditionally conclude that the state of V = 0
  ```

### 4.1.3 THE INTERACTIVE TEXT EDITOR

The interactive *text editor* in G2 is used for editing text in statements, rules, functions, and so on. It operates through a *text-edit workspace* that appears on the screen when the developer starts to edit text. Within this workspace, lists are highlighted, indicating the options for the next possible phrases. For example, when editing a rule, the text editor lists the possible first words. As can be seen in Fig. 10.7 the text editor even lists the names of the items in the knowledge base and the developer may choose from this list or may enter the text by typing on the keyboard. G2 marks syntactically incorrect text with an ellipsis and displays a message below it, only accepting syntactically correct texts.

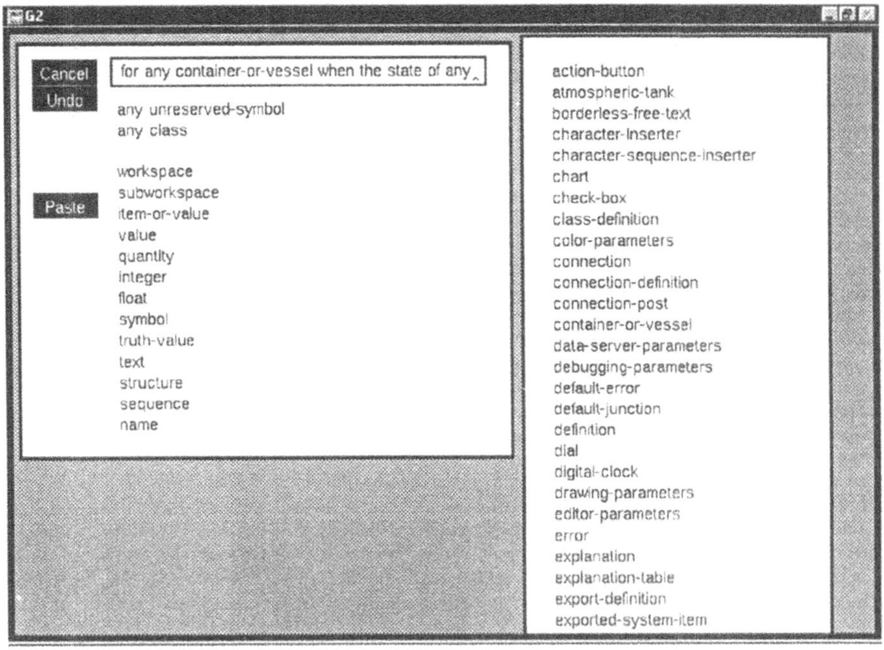


*Figure 10.7.* Interactive text editor

### 4.1.4 THE INTERACTIVE ICON EDITOR

The interactive *icon editor* helps to create and modify icons with graphic tools and convert the graphic description into G2 grammar. An icon consists of one or more overlapping *layers*, which are transparent films with single-colored pictures. The layers can be grouped into *regions* and all of the layers in a region have the same colors.

As can be seen in Fig. 10.8 the icon editor has several important parts:

- the *icon view* box shows what the icon looks like

- graphic *buttons* are used to create graphic elements, to undo and complete actions and expand the view

- the *icon size display* shows the size of an icon in terms of workspace units

- the *cursor location display* gives the exact location of the mouse pointer in terms of coordinates

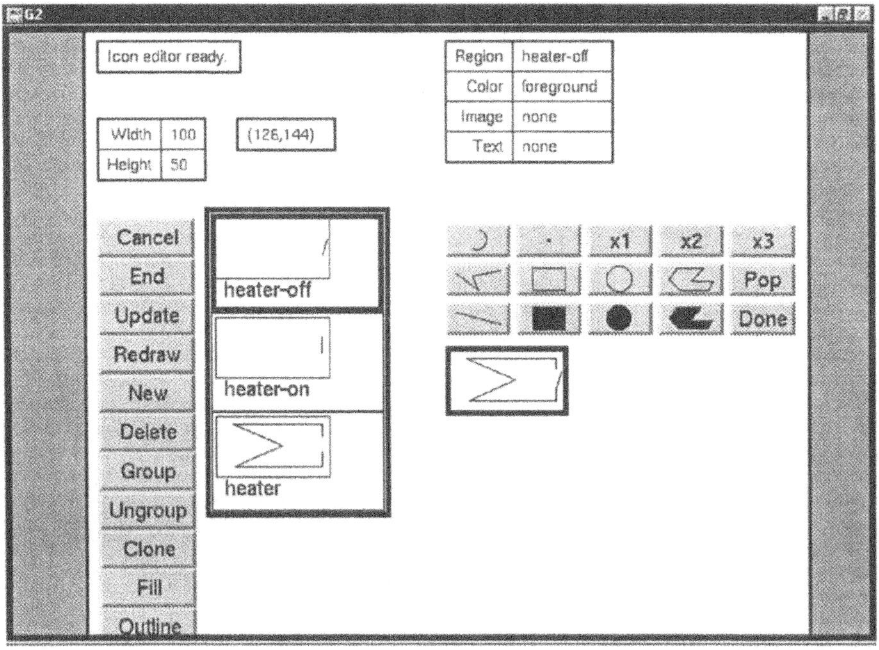


*Figure 10.8.* **Interactive icon editor**

- the ***layer pad*** shows the layers of an icon. Layers can be added, deleted, grouped together, assigned with region labels and colors, etc. A heavy border indicates layer which is currently being edited.

### 4.1.5 KNOWLEDGE BASE HANDLING TOOLS

G2 has several knowledge base handling tools which are used to produce, modify and run a large and complex knowledge base. These tools are:

- ***cloning*** items helps the creation of similar items easily. This makes it possible to build a large knowledge base quickly.

- ***carrying out an operation on a group of objects*** helps to avoid performing the same function more than once.

- ***inspecting*** a knowledge base (as in Fig. 10.9) makes it easy to find items and to browse a large knowledge base quickly.

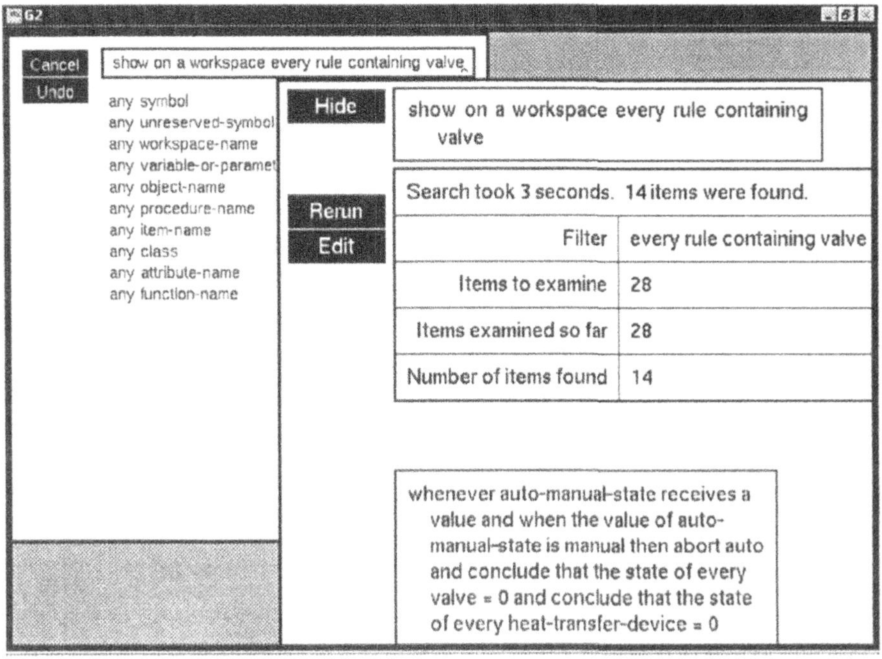


*Figure 10.9.* **Inspecting knowledge base**

- *describing* variables (as in Fig. 10.10) specifies the data server corresponding to the variable and the rules according to which the variable receives values.

- *hierarchical organization of the knowledge base* makes it easier to understand and use the knowledge base.

- *merging the knowledge base* is a tool used to create one knowledge base from two.

### 4.1.6 DOCUMENTING IN THE KNOWLEDGE BASE

*Free texts* can be attached to workspaces in G2 applications. Free texts don't affect the knowledge base, but only document it. The developer can define document objects, which have subworkspaces with free texts containing information.

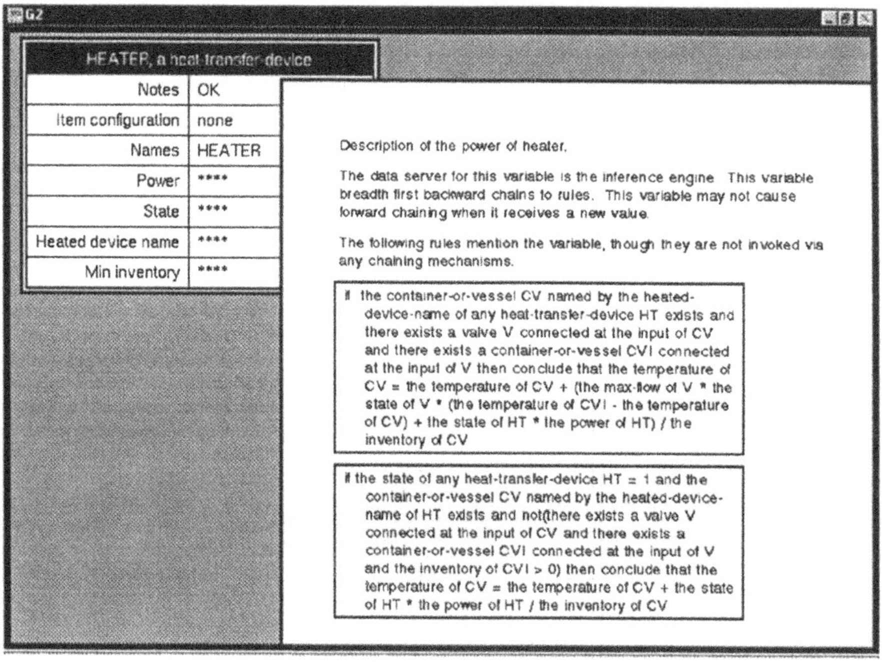


*Figure 10.10.* **Describing variables and parameters**

### 4.1.7 TRACING AND DEBUGGING FACILITIES

G2 gives dynamic feedback to the developer when it invokes rules, executes formulas, functions, procedures or variables. G2 has the followig debugging and tracing facilities:

- displaying ***warning messages*** about errors and unexpected events
- displaying ***trace messages*** that show:
  - the current value of a variable or expression whenever it receives a new value
  - the time when G2 starts and stops the evaluation of a variable, rule, formula, procedure or function
  - the time when G2 executes each step in the evaluation process
- generating ***breakpoints*** at each step of the evaluation process
- ***highlighting*** invoked rules

Warning and trace messages may apply to the whole knowledge base or certain parts of it.

### 4.1.8 THE ACCESS CONTROL FACILITY

The *access control* in G2 is used to control what different user groups can see and do within a knowledge base. The access control facilities are as follows:

- limiting the number of menu-options available to a user
- preventing users from for example moving, connecting, cloning items
- allowing users to see only part of an attribute table
- allowing users to see the attributes of an item without editing them or creating a subworkspace, etc.

These restrictions may be applied to all items in the knowledge base, to certain classes of items, to the items on a certain workspace, or to individual items. Several user modes or groups (for example operator, administrator, developer) may be defined by the developer by setting different access controls.

## 4.2 THE END-USER INTERFACE

There are several tools that aid communication between G2 and a user. Some of them are described in section 4.1 of this Chapter. G2 also provides a number of predefined objects, which inform end-users about the status of the knowledge base when it's running. These include:

- *displays*, which show the values of variables, parameters or expressions
- *end-user controls*
- *messages, message board and a logbook* as tools for communicating with the end-user

### 4.2.1 DISPLAYS

*Displays* are devices that show the user the values of a variable or expression. G2 provides five types of displays:

- a *readout table* is a box that shows a variable, parameter or expression and its value.
- a *chart* plots the values of one or more numeric expressions over time.
- a *meter* shows the value of an arithmetic expression as a vertical bar along a numeric scale.
- a *dial* shows the value of an arithmetic expression as a pointer that rotates along a circular numeric scale.

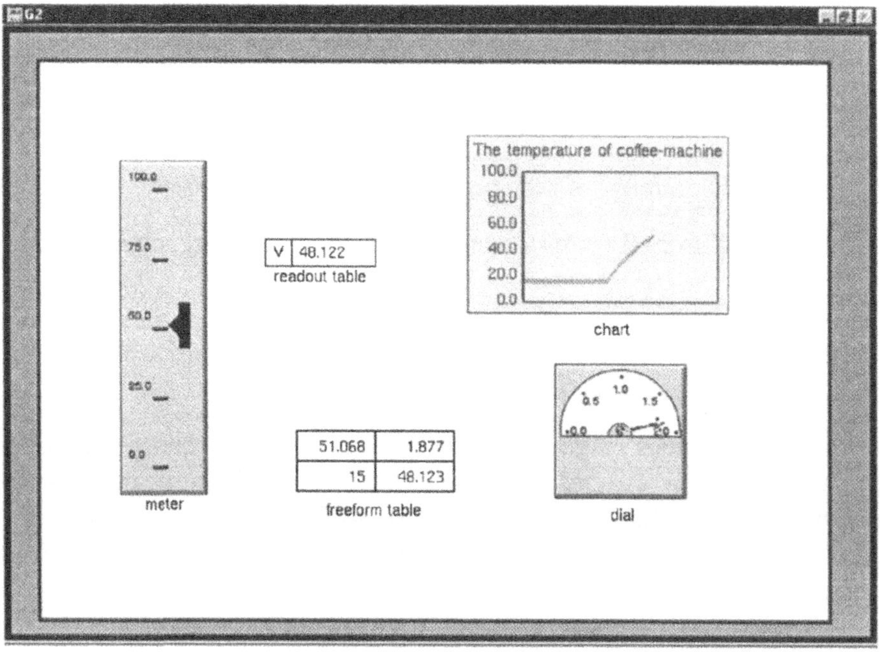


*Figure 10.11.* Displays

- a *free-form table* displays values of variables or expressions in cells arranged in rows and columns.

An example of every display type is shown in Fig. 10.11.

### 4.2.2 END-USER CONTROLS

*End-user controls* are devices that end-user can use to control an application. As Fig. 10.12 shows, there are five kinds of end-user controls:

- an *action button* is a rounded, rectangular box, which causes G2 to execute one or more actions like `start`, `conclude`, `show`, and so on, when a user clicks on it.

- a *radio button* is used to assign a predefined symbol, number, text, or logical value to a variable when a user clicks on it. It is a small circle in which a black dot appears when it is selected.

- a *check box* is a small, square box, which assigns an "on" or "off" value to a variable when the user clicks on it.

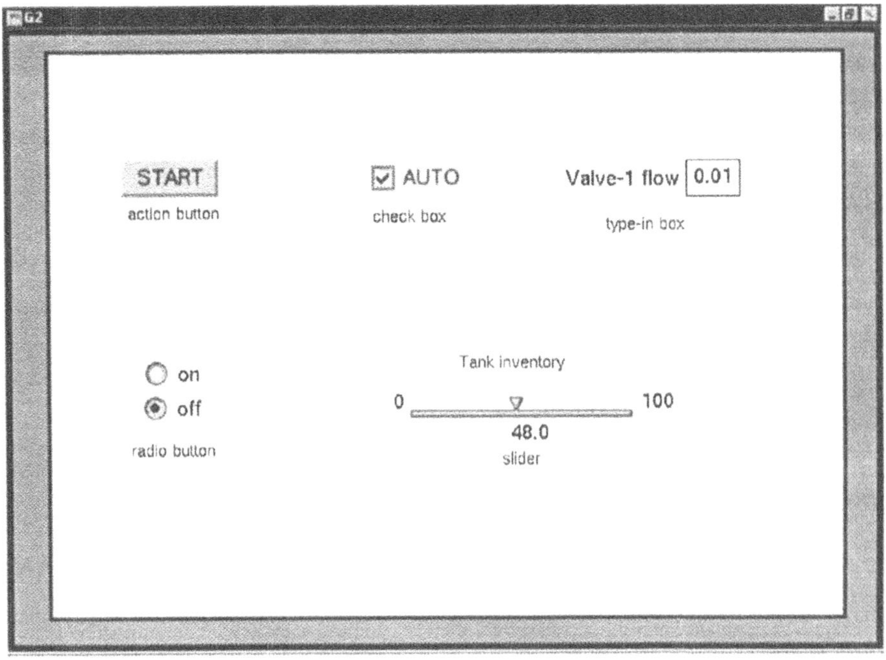


*Figure 10.12.* **End-user controls**

- a ***slider*** is a horizontal line with numbers at either end, allowing a user to enter numeric values by sliding a pointer to the appropriate position.

- a ***type-in box*** is used for entering values using the keyboard.

### 4.2.3 MESSAGES, MESSAGE BOARD AND LOGBOOK

A ***message*** is an item that displays text. G2 may inform the user by showing messages on the ***message board*** or in the ***logbook***. Messages which appear as a result of `inform` action are instances of the built-in message class. The developer can define subclasses of the message class with their specific attributes and characteristics.

The message table and the message board are two workspaces where messages may appear. Messages generated by an `inform` action in rules generally appear on the message board or in any workspace. G2 writes its messages in the logbook about system conditions, errors and warnings.

## 4.3 EXTERNAL INTERFACE

G2 has several interfaces, which support interaction with other processes and the receiving of data from external sources. These are easy to configure and, because they work automatically while a knowledge base is running, easy to use. The interfaces available for use with G2 are as follows:

- *G2 Standard Interface (GSI)* helps building interfaces between G2 and external processes and systems
- *G2 File Interface (GFI)* enables G2 to write or read data files
- *G2 Simulator Interface (GSPAN)* may attach G2 to an external simulator
- *G2-G2 Interface* enables two G2s to communicate
- *Foreign Function Interface* supports the calling of C or FORTRAN functions in G2

# Appendix A
# A BRIEF OVERVIEW OF COMPUTER CONTROLLED SYSTEMS

Computer controlled systems are basic components in almost every intelligent control system. Therefore the basic concepts, notions and techniques of computer controlled systems are needed to understand the material in this book.

All the material that is not included in standard engineering curriculum, namely fundamentals of systems and control theory as well as software engineering of real-time control systems is summarized in this chapter. The material is divided into the following sections.

- Basic notions in systems and control theory [92], [93]

- State-space models of linear and nonlinear systems [93], [94]

- Common functions of a computer controlled system [93]

- Real-time software systems [95]

- Software elements of computer controlled systems [93]

## 1. BASIC NOTIONS IN SYSTEMS AND CONTROL THEORY

Systems and control theory is a well grounded engineering discipline with rigorous mathematical background [92], [93], [94]. It relies on two fundamental concepts: on the concept of signals and signal spaces and that of systems.

## 1.1 SIGNALS AND SIGNAL SPACES

Real-world objects with time-dependent behaviour act on each other in various ways. We describe these interactions using scalar- or vector-valued time-dependent functions, which are called *signals*.

If we consider a vector-valued signal

$$x : \mathcal{R} \to \mathcal{R}^n$$

then the value of this signal at any given time instance $t$, $x(t)$ is a vector. Sometimes the value of a signal at a given time instance can be a space-dependent function.

The set of all possible time-dependent functions which can be realizations of a signal form a *signal space* $\mathcal{X}$ associated with the signal $x$.

## 1.2 SYSTEMS

We understand the *system* to be part of the real word with a boundary between it and its environment. The system interacts with its environment only through its boundary. The effects of the environment on the system are described by time dependent ***input*** functions $u(t)$ from a given set of possible inputs $u \in \mathcal{U}$ , while the effect of the system on its environment is described by the ***output*** functions $y(t)$ taken from a set of possible outputs $y \in \mathcal{Y}$. The schematic ***signal flow diagram*** of a system **S** with its input and output signals is shown in Fig. A.1.

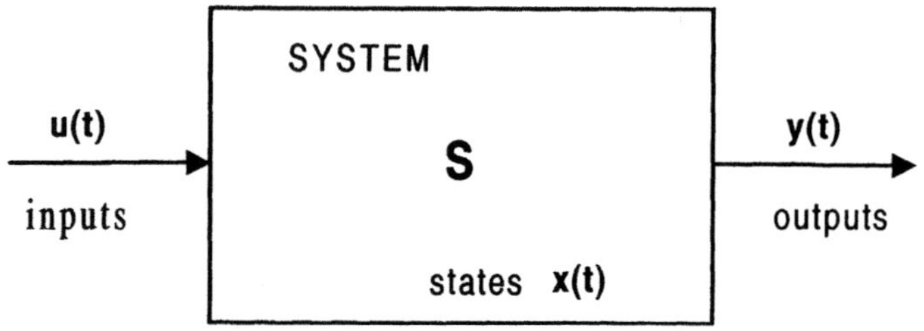


*Figure A.1.* Signal flow diagram of a system

We can look at the signals of a system as the input causing its time dependent behaviour that we can observe in its output.

There are systems which have especially interesting properties and are easy to handle from the viewpoint of their analysis and control.

- *linearity*

  The first property of special interest is linearity. A system **S** is called *linear* if it responds to a linear combination of its possible input functions with the same linear combination of the corresponding output functions. Thus for the linear system we note that:

$$\mathbf{S}[c_1u_1 + c_2u_2] = c_1y_1 + c_2y_2 \tag{A.1}$$

with $c_1, c_2 \in \mathcal{R}$, $u_1, u_2 \in \mathcal{U}$, $y_1, y_2 \in \mathcal{Y}$ and $\mathbf{S}[u_1] = y_1$ , $\mathbf{S}[u_2] = y_2$.

- *time-invariance*
  The second interesting class of systems are time-invariant systems. A system $\mathbf{S}$ is *time-invariant* if its response to a given input is invariant under time shifting. Loosely speaking, time-invariant systems do not change their system properties in time. If we were to repeat an experiment under the same circumstances at some later time we would get the same response.

  The system parameters of a time-invariant system are constants, i.e. they do not depend on time.

- *continuous and discrete time systems*
  We may classify systems according to the time variable $t \in \mathcal{T}$ we apply to their description. There are *continuous time* systems where time is an open interval of the real line ($\mathcal{T} \subseteq \mathcal{R}$). Discrete time systems have an ordered set $\mathcal{T} = \{\cdots, t_0, t_1, t_2, \cdots\}$ as their time variable set.

- *single-input single-output (SISO) and multiple-input multiple-output (MIMO) systems*
  Here the classification is determined by the number of input and output variables.

## 2. STATE-SPACE MODELS OF LINEAR AND NONLINEAR SYSTEMS

In the most general and abstract case we describe a system by an *operator* $\mathbf{S}$. However, in most of the practical cases outlined in the subsequent subsections we give a particular form of this operator. The operator $\mathbf{S}$ can also be characterized by a set of parameters $p$, which are called *system parameters*.

In order to obtain the so called state-space description [92], [93], [94]. let us introduce a new variable, called the *state of the system at* $t_0$, which contains all past information on the system up to time $t_0$. Then for *causal systems* we only need $u(t)$, $t \geq t_0$ and the state at $t = t_0$ to compute $y(t)$ for $t \geq t_0$ (all future values). If the state of a nonlinear system can be described at any time instance by a finite dimensional vector then the system is called a *concentrated parameter system.*

## 2.1 STATE-SPACE MODELS OF LINEAR TIME-INVARIANT SYSTEMS

It can be shown that *the general form of the state-space representation or the state-space model* of multi-input multi-output (MIMO) linear time-invariant (LTI) systems is as follows: :

$$\begin{array}{ll} \dot{x}(t) = Ax(t) + Bu(t) & (\textit{state equation}) \\ y(t) = Cx(t) + Du(t) & (\textit{output equation}) \end{array} \tag{A.2}$$

with the initial condition $x(t_0) = x(0)$ and

$$x(t) \in \mathcal{R}^n \ , \ y(t) \in \mathcal{R}^m \ , \ u(t) \in \mathcal{R}^r \tag{A.3}$$

being vectors of finite dimensional spaces and

$$A \in \mathcal{R}^{n \times n} \ , \ B \in \mathcal{R}^{n \times r} \ , \ C \in \mathcal{R}^{m \times n} \ , \ D \in \mathcal{R}^{m \times r} \tag{A.4}$$

being matrices. Note that $A$ is called the *state matrix*, $B$ is the *input matrix*, $C$ is the *output matrix* and $D$ is the *input-to-output coupling matrix*.

The parameters of a state-space model consist of the constant matrices

$$p = \{ \ A, \ B, \ C, \ D\}$$

The *state-space representation (SSR) of an LTI system* is the quadruplet of the constant matrices $(A, B, C, D)$ in equation (A.2). *The dimension of an SSR* is the dimension of the state vector: $dim \ x(t) = n$. State-space $\mathcal{X}$ is the set of all states:

$$x(t) \in \mathcal{X} \quad , \quad dim \ \mathcal{X} = n \tag{A.5}$$

## 2.2 STATE-SPACE MODELS OF NONLINEAR TIME-INVARIANT SYSTEMS

Having identified the relevant input, output, state and disturbance variables for a concentrated parameter nonlinear system, the *general nonlinear state-space equations* can be written in matrix form:

$$\frac{dx(t)}{dt} = \mathbf{f}(x(t), u(t), p) \tag{A.6}$$

$$y(t) = \mathbf{h}(x(t), u(t), p) \tag{A.7}$$

The nonlinear vector-vector functions $\mathbf{f}$ and $\mathbf{h}$ in equation (A.7) characterize the nonlinear system. Their parameters constitute the system parameters.

## 2.3 CONTROLLABILITY

Conditions to check controllability will be given for LTI systems with finite dimensional representations in the form

$$\begin{aligned} \dot{x} &= Ax + Bu \\ y &= Cx \end{aligned} \tag{A.8}$$

Observe that from now on we assume $D = 0$ in the general form of the state-space representation (abbreviated as SSR) in equation (A.8). Therefore, an SSR will be characterized by the triplet $(A, B, C)$. Note also that state-space representations are not unique: there is an infinite number of equivalent state-space representations giving rise to the same input-output behaviour.

A system is called *(state) controllable* if we can always find an appropriate manipulable input function which moves the system from its given initial state to a specified final state in finite time. This applies to every given initial state final state pair.

The problem statement for state controllability can be formalized as follows.

STATE CONTROLLABILITY

*Given:*
The state-space representation form with its parameters as in Eq. (A.8) and the initial $x(t_1)$ and final $x(t_2) \neq x(t_1)$ states respectively.

*Question:*
Is it possible to drive the system from $x(t_1)$ to $x(t_2)$ in finite time?

•

For LTI systems there is a necessary and sufficient condition for state controllability which is stated in the following theorem.

**Theorem A.1.** *An SSR $(A, B, C)$ is state controllable if and only if the controllability matrix $\mathcal{C}_n$:*

$$\mathcal{C}_n = \begin{bmatrix} B & AB & \cdots & A^{n-1}B \end{bmatrix} \tag{A.9}$$

*is of full rank, that is*

$$rank\ \mathcal{C}_n = n$$

Note that *controllability is a realization property*, and it may change if we apply state transformations to the state-space representation.

## 2.4 OBSERVABILITY

The notion of observability originates from the fact that the states of a system are assumed not to be directly measurable. We can only measure directly the input and output signals and then compute or estimate the value of the state variables. This cannot be done in all cases, only when the observability property holds.

A system is called *(state) observable* if we can compute the value of the state variables at a given time instance, say at $t = t_0$ from a finite measurement record of the input and output variables and from the system model. The problem statement for state observability is again given in the form of a problem statement.

STATE OBSERVABILITY

*Given:*
The state-space representation form with its parameters as in Eq. (A.8) and finite measurement records for the input and output variables in the form of

$$\{u(t) \mid t_0 \leq t \leq T\} \quad , \quad \{y(t) \mid t_0 \leq t \leq T\}$$

respectively.

*Question:*
Is it possible to compute the value of the state variable at $t = t_0$, that is to determine $x(t_0)$?

•

For LTI systems there is a necessary and sufficient condition for state observability, which is stated below.

**Theorem A.2.** *An SSR $(A, B, C)$ is state observable if and only if the observability matrix $\mathcal{O}_n$:*

$$\mathcal{O}_n = \begin{bmatrix} C \\ CA \\ \cdots \\ CA^{n-1} \end{bmatrix} \tag{A.10}$$

*is of full rank, that is*

$$rank\ \mathcal{O}_n = n$$

Observe that observability only depends on the matrices $(A, C)$ but not on $B$. *Note that observability is a dual property of controllability and it is also a realization property.*

## 2.5 STABILITY

Stability characterizes how a given system reacts to disturbances. There are two basically different stability notions:

- *bounded-input bounded-output (BIBO) stability*
  describes what happens if the system receives a bounded input signal.

  The system may respond with a bounded output signal to any bounded input signal, in which case we call it a BIBO stable system.

- *asymptotic stability*
  tells us what happens if we move the system from its equilibrium or steady-state and then leave it alone.

  If the perturbed system goes back to its original steady state after a long time (i.e. asymptotically) then we call the system asymptotically stable.

Both asymptotic and BIBO stability are system properties for LTI systems, where asymptotic stability implies BIBO stability. The problem statement for asymptotic stability in the case of LTI systems is given below.

ASYMPTOTIC STABILITY

*Given:*
The state equation of the state-space representation form as in Eq. (A.8 but with zero input, i.e. $u(t) \equiv 0$ and with nonzero initial condition:

$$\dot{x} = Ax \quad , \quad x(0) = x_0 \neq 0$$

*Question:*
Will $x(t)$ go to zero in the limit i.e.

$$\lim_{t \to \infty} x(t) = 0 \ ?$$

There is a simple necessary and sufficient condition for a LTI system to be asymptotically stable, which is stated by the following theorem.

**Theorem A.3.** *A LTI system with a state matrix $A$ is asymptotically stable if and only if the real parts of all the eigenvalues of the state matrix are strictly negative, that is*

$$Re\{\lambda_{Ai}\} \ < \ 0 \quad , \quad i = 1, ..., n$$

Observe that asymptotic stability only depends on the state matrix $A$ but not on the other two matrices in a SSR, i.e. on $B$ and $C$.

# 3. COMMON FUNCTIONS OF A COMPUTER CONTROLLED SYSTEM

Although the software architecture of computer controlled systems may vary widely with the application for which they are designed, there are characteristic software components present in each of them [93]. In order to investigate these we first need to briefly review the common functions of computer controlled systems and then have a look at the common features of real-time software systems.

Almost any computer controlled system has two data sources and several targets in its environment:

- the *plant or process* to be controlled,
- the *users* of various kinds (engineers, operating personnel etc.)

The common and specific functions of computer controlled systems mainly belong to the functions of the *computer-plant interface*. They can be classified into the following groups according to the level of abstraction and the direction of data transfer.

1. *primary/secondary data processing functions*
2. *process monitoring functions*
3. *process control functions*

The functions in these groups are described in detail in the following subsections.

## 3.1 PRIMARY DATA PROCESSING

Sensors and other measurement devices produce unscaled signals together with coded status information on the state of the measurement device. These measured (signal-status) pairs are the so called raw measured data. The aim of primary processing is to produce scaled, validated and verified data which can be used in engineering context from raw measured data (also called (primary) measured data). It is important that raw measured data coming from a particular sensor form a time dependent sequence, that is, a discrete time signal from the point of view of system theory.

Secondary data processing carries out more sophisticated data analysis and verification procedures applied to measured data.

The primary functions of data acquisition and data analysis belong to this group, which can be further classified into the following subfunctions.

- *handling missing or invalid data*
  This usually involves checking the status information of raw measured data for sensor failure or malfunction. In such situations the obtained value is invalidated. If needed, invalid data are substituted with previous valid values.

- *scaling*
  Scaling is one of the most important primary processing steps from the users' point of view. With the help of equipment scaling and calibration data, a raw value is transformed into a scaled value in engineering units.

- *limit checking*
  Most measurement devices have a measurement range associated with them and there is a signal in their status information when the raw measured value is found to be outside this range. These limits are considered as "hard" limits.

  The underlying technology usually determines narrower range(s), so called "soft limits" within which a particular measured value should be. Most often two sets of upper and lower limits are considered: the *warning limits* and the *error limits.* The upper and lower limit values are a'priori given static data, which are stored within the set of *primary processing data.*

  Limit checking is then usually performed by a simple arithmetic comparison of measured data and limits.

- *filtering*
  The aim of the filtering sub-step in primary or secondary processing is to remove outing values and reduce the variation in measured data by using simple on-line methods.

  The removal of outlying values is performed by limit checking and data removal or substitution. Simple signal filtering methods such as weighted averaging, averaging with exponential filtering or 1st order linear filters with constant coefficients are used here.

  The necessary parameters and filter coefficients are stored in *primary processing data.*

- *averaging*
  A set of time-dependent measured data sequences is averaged for different reasons. Averaging is used as a simple signal filtering method (see above), but there are averages over a longer operation period, say over a shift, day or month, which are used for monitoring purposes.

Averaging can be performed recursively in an on-line manner, when only the current average and newly measured data are required to calculate an updated version of the average.

## 3.2 PROCESS MONITORING FUNCTIONS

This group of functions aims at informing the operators about the status and performance of the plant to be controlled, and about the status of measurement devices and actuators.

The functions use measured data produced by primary/secondary data processing, that is they work on scaled and validated measured data in engineering units.

- *alarm generation*

  As a result of limit checking and the detection of missing or invalid data and outlying values, various warning and error messages are generated. These messages are presented to system operator(s) and are also stored as events by the computer controlled system itself.

  Some alarm messages require actions from operator(s), for example manual acknowledgement of the message.

- *computation of process trends*

  Process trends describe the time-variation of a measured data signal or a group of signals in order to discover and detect drifts and periodic changes in the value of the signal over a long operating range. Process trends are usually presented on a plot and detected by fitting a curve on measured data signals.

  The computation of process trends may require filtered or short-term averaged data. Consequently, it is closely related to secondary data processing.

- *logsheet generation*

  A logsheet is a pre-arranged condensed set of information for a given operational or maintenance purpose, produced periodically in each prescribed time interval (say daily) or upon request. A logsheet usually contains complex data such as averages, filtered data or trends. Various statistics, such as histograms of data values are also often included.

## 3.3 PROCESS CONTROL FUNCTIONS

The aim of process control functions is to influence the behaviour of the plant to be controlled in order to achieve some prescribed goal. Thus these functions are most often active functions in the sense that they produce signals which influence the plant. These signal values are stored in the set of *actuator data.* and usually computed from the set of measured data.

Besides the active control or regulation sub-function, process control functions most often include preparatory or auxiliary functions for control, such as filtering, identification or diagnosis.

- *control and regulation*
  Controllers of various kinds are applied to achieve a specific aim with respect to the plant, such as moving it from one operating point to another or keeping it at an operating point despite the effect of disturbances. Regulation is a special case of control when we want to keep a signal or a group of signals constant.

  Using the measured past and present input and output signal values of a system, controllers compute the actual value of the input signal that is used to influence the system. Thus control functions are typically active functions in a computer controlled system, which determine the value of system actuators.

  The most common regulator is the so called *PID controller.*

- *state filtering*
  A big group of controllers, for example LQRs or pole placement controllers, apply state feedback to the system, that is they use the value of present state signals to compute control input. As state signals are not directly measurable and we only have measured data available, which is corrupted by measurement noise, we need to perform state filtering to obtain an estimate of the state signal values. The most famous state filtering method is the *Kalman filter.*

- *identification*
  Control methods require a complete dynamic model of a system including the value of its parameters. These system parameters are usually not precisely known and may also vary in time. Therefore we need to apply identification methods to determine system parameters from system structure and measured data.

- *diagnosis*
  Diagnosis aims to discover, detect and isolate plant faults and malfunction from the measured data and from model of the "healthy" and "non-healthy" plant in different faulty modes. It provides advanced information for the operators on the state of the plant and also guides the operation of controllers.

### 3.4 FUNCTIONAL DESIGN REQUIREMENTS

The common functions of computer controlled systems require the presence of certain functions and properties in the software system, which is used to implement them. Some of the requirements follow from the time-dependent nature of the system to be controlled, others are in connection with the technical or algorithmic nature of the tasks to be performed.

Functional design implies that of computer controlled systems as software systems need to have the following important characteristics:

- *handling of time dependence*
  This requirement follows from the time-dependent nature of systems and controllers.

- *handling of measurement devices and actuations*
  The input and output signals of a system are measured quantities varying in time, which calls for the presence of measurement devices (sensors). Actuators are needed to implement control functions.

- *handling of events*
  An event is a discrete change in a system at a given time instance. Any warning or error message as well as the actions of operator(s) or controllers are regarded as events.

These characteristics make it necessary to use a *real-time software system* as an implementation environment for computer controlled systems.

## 4. REAL-TIME SOFTWARE SYSTEMS

Real-time software systems are briefly described in this section in order to show that they possess the properties necessary for the implementation of computer controlled systems. Special emphasis is put on those characteristics, tools and elements influencing the architecture of real-time expert systems, to which intelligent control systems belong. More about real-time systems can be found elsewhere e.g. in [95].

### 4.1 CHARACTERISTICS OF REAL-TIME SOFTWARE SYSTEMS

A real-time software system should be able to react to randomly occurring events and perform time-dependent tasks. Moreover, in a real industrial environment it should operate under highly varying load, when the number of signal changes may vary widely according to whether the system is in the quiet "nothing happening" situation to the hectic "full system alarm" status.

Therefore, a real-time operating system should have the following properties in the form of standard operating system service functions:

- *real-time clock*
  A real-time software system should have an independent central element, a clock, which operates independently of the load and circumstances. All time-dependent functions and services use the value given by the system clock.

- *handling time*
  The presence of the clock makes it possible to handle timed tasks, such as periodic tasks or tasks to be performed at a given time instance.

- *time dependent behaviour*
  Most often there is a need to follow control sequences, that is timed sequences of prescribed actions within a computer controlled system. These control sequences may perform operations on the system to be controlled and may also influence the state and operation of the software system itself.

- *event handling*
  The behaviour of the environment - that is, the system to be controlled - and the users constitute events that influence the computer controlled system. An event describes a specific change that occurred at a specific time instance in the abstract form of a

  $$(change\_identifier,\ time\_stamp)$$

  pair.

- *priority handling*
  In real circumstances, the load of a computer controlled system, which can be measured by the number of signal changes, varies widely in time. At the same time, computer controlled systems are usually designed for an average load. consequently, the system is highly overloaded from time-to-time. In such situation, the system should focus on the most important tasks and omit or delay tasks of secondary importance.

  Priority handling is a technique to ensure the *nice degradation* of system performance by defining priority classes, allocating a priority to each task and executing them in the order of priority.

## 4.2 ELEMENTS OF REAL-TIME SOFTWARE SYSTEMS

The architecture of a real-time system is described in terms of its elements and their connections. The elements of any software system can be categorized using Wirth's formula:

*"programs = data structures + algorithms"*

The elements of real-time software systems are then categorized as follows.

1. *tasks (processes)*
   these are the active elements of a real-time system implementing the "algorithms". Typically, there are a number of autonomous and relatively independent tasks in a real-time system.

2. *data files*
   The data structures in a real-time system are described by data files and are collected in a real-time database.

3. *interfaces*
   Interfaces are special active elements dealing with resource allocation, organization and synchronization of and communication between the elements of a real-time system and its environment. Based on the elements the interface connects, the following interface categories are distinguished.

   - task-task interface
   - task-file interface
   - human-computer interface

These elements and their connections are the subject of software design in a real-time system.

## 4.3 TASKS IN A REAL-TIME SYSTEM

Here we briefly summarize the general properties of tasks and their interfaces in a real-time software system. More can be found about the tasks in a computer controlled system in the next section.

1. *Task states and state transitions*
   Any task in a real-time operating system may exist in standardized states depending on its position in its life-cycle and the status of its environment. Task states and state transitions are administered by the *scheduler* of the operating system, which is a special high-priority task with scheduling and resource allocation capabilities.

The task states and state transitions are depicted in Fig. A.2 in the form of a state transition diagram borrowed from the theory of discrete automata.

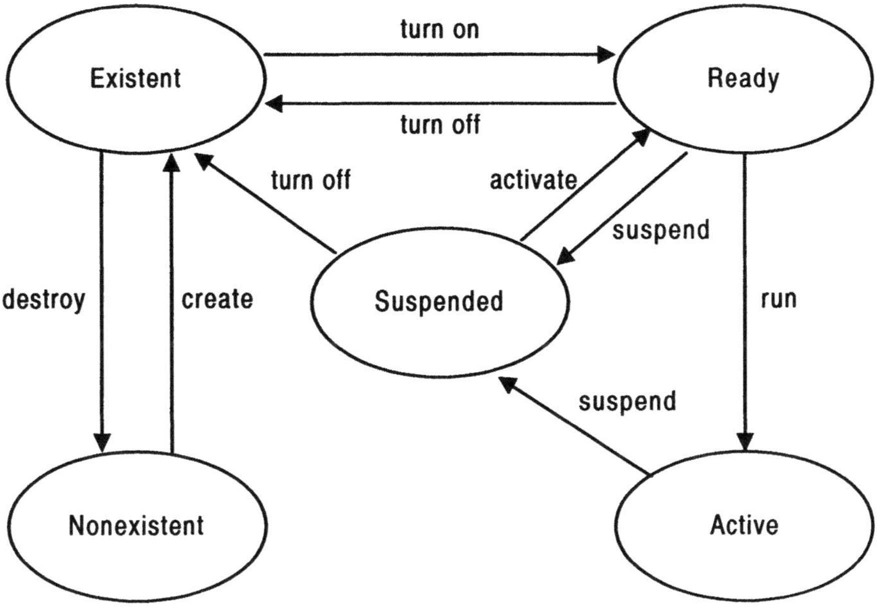


*Figure A.2.* Tasks states and state transitions

2. *Task-task interfaces*
   The organization and functions of task-task interfaces are also standardized in their form and primitives. There are two types of task-task interfaces: the *synchronization* interface only deals with the timing and synchronizing of task execution, while the *communication* type interface allows the exchange of data structures together with synchronization.

   - synchronization
     There are two types of synchronization between two tasks: the one-way *go* and the two-way *randezvous.*
     The usual way of implementing synchronization interfaces is to use system flags, one for the *go* and two for the *randezvous* type connection. The necessary communication primitives for implementing a synchronization interface are
     - set-flag
     - wait-for-flag

- communication
  Similarly to synchronization connections, two types of communication connections exist: the one-way *send* and the two-way *send-and receive* connection. They can be implemented using

  - database files and flags
  - mailboxes (queues: FIFO, LIFO)

  in a real-time operating system.

3. *Scheduler*
   As we have seen before, the scheduler is a special high priority task in a real-time system dealing with task states and state transitions. Besides this, the ***scheduler has other duties*** in a real-time software system. These are the following.

   - interrupt handling and administering
   - clock management (sometimes this is a special task in itself)
   - providing an interface for database management
   - providing an interface for measurement handling
   - management of timing
   - management of task-task synchronization
   - management of task-task communication

4. *Operation of tasks*
   Tasks in a real-time system usually perform a sequence of synchronization or communication operations together with algorithmic data processing operations.

   Tasks in a computer controlled system perform a typical cyclic operation the following way. After an initialization sequence which is executed once when the task changes its state from the "Existent" to "Ready" for the first time, a cycle of operation is executed every time when the task moves from "Suspended" to "Ready" and back. This is illustrated by a typical task frame in Example A.1.

**Example A.1.** *A typical task frame in computer controlled systems*

The following program frame shows a typical task frame in a computer controlled system in Pidgin Algol syntax. It uses the *randezvous* type synchronization between the task and its task mates. Observe that two flags, **flag1** and **flag2** are needed to implement this connection.

```
      initialization;
loop: wait-for-flag(flag1);     { waiting for task starting }
      get-message-or-data;      { real operation starts }
      process data;
      put-message-or-data;      { real operation ends }
      set-flag(flag2);          { signalizing the ready state }
      goto loop;
```

Finally it is important to note, that there are *typical problems* inherent in real-time systems, which are as follows.

- *the danger of dead-locks*
  If resource allocation rules and their management system is poorly designed, a dead-lock situation may arise. This happens when tasks are allowed to request their resources (flags, database files etc.) in a sequential incremental manner and then a group of tasks may wait for each other to get the requested resources released.

- *consistency management of database files*
  Real-time software systems need a special *real-time database management system* to take care of the time-dependent values of measured signals and actuators, as well as that of events. It is important to ensure that data files are consistent at a given time instance. The ability to lock a record and a whole data file may be necessary for this purpose. Therefore any real-time database management system has an advanced resource management and archiving system as compared to conventional database management systems.

- *"graceful degradation" property*
  As it has been mentioned before, real-time systems often operate with widely varying load, which can be high compared to the average load they have been designed for. Graceful degradation means that there are tools and techniques to perform the necessary, most important

tasks with a high priority and delay or even omit the less important tasks.

# 5. SOFTWARE ELEMENTS OF COMPUTER CONTROLLED SYSTEMS

Computer controlled systems are special real-time software systems, which have typical data structures (or data files) and tasks. The most important software elements, tasks and data structures are briefly described here [93].

The connection between tasks and data files in a computer controlled system are shown in Fig. A.3. The solid arrows denote read/write connections and the dashed arrows denote synchronization connections.

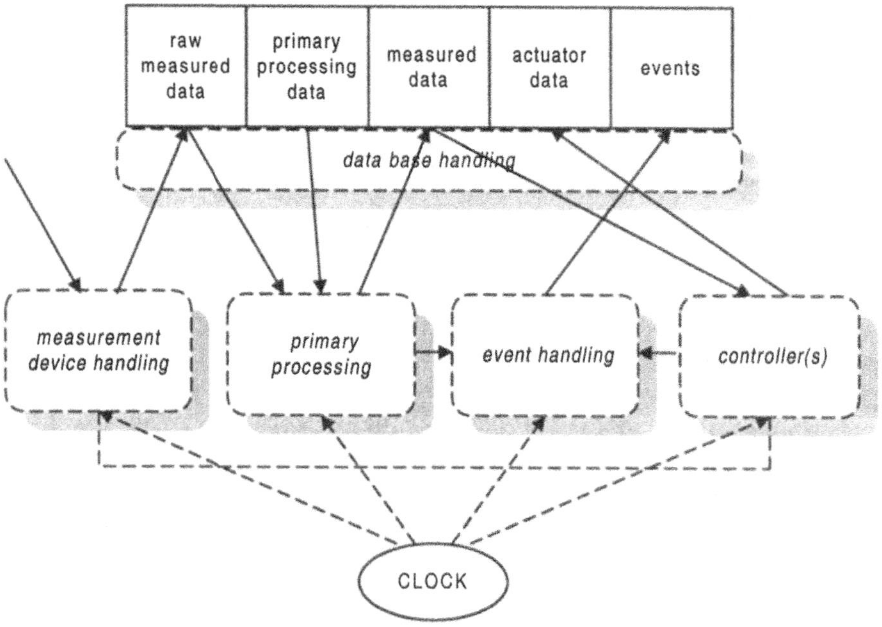


*Figure A.3.* Typical tasks and data files of CCSs

## 5.1 CHARACTERISTIC DATA STRUCTURES OF COMPUTER CONTROLLED SYSTEMS

Typical data structures are used to store the ingredients of measured data and events needed for the operation of computer controlled systems. The following characteristic data files can be distinguished:

1. *raw measured data* and *measured data* files
2. *primary processing data* file
3. *events* file
4. *actuator data file*

The data files above are briefly described below.

### 5.1.1 RAW MEASURED DATA AND MEASURED DATA FILES

The *raw measured data* file is generated by the *measurement device handling* task and contains the primary results received by the plant sensors. Remember that sensors do not only send the unscaled raw value of the quantity they measure but also provide status information.

The *measured data* file is then filled by the *primary processing* task with the scaled and validated measured data. This file contains the results of primary processing and serves as a basic data source for all the other processing functions, such as secondary processing and control tasks.

Both files contain the fields

- `measurement device identifier`
  The measurement device identifier is a unique name which refers to both the signal this record belongs to and the measurement device type.

- `measured data`
  This field contain the most important information in this file. The value is unscaled for the raw data. The length of this field varies with the type (real or binary) of the signal it belongs to.

- `status`
  For raw measured data, status information is directly sent by the corresponding sensor with information on the status of the raw measured value, which can be {*non–valid*, *measurement limits exceeded*, *time–out*}, etc.

  Primary processing adds more information to measurement status by indicating if the raw measured value exceeded a warning or alarm limit, or has been found to be an outlying value.

- `time stamp`
  Sensors send values when they substantially change, that is at irregular time intervals. The `time stamp` field in a record tells us the time

instance when the value was last updated thus providing information on the change of the value and on its validity.

### 5.1.2 PRIMARY PROCESSING DATA FILE

This is a constant data file used by the *primary processing task* to perform primary and secondary processing functions on raw measured data. It contains the following time-independent information on sensors and measured variables.

- `measurement device identifier`
  The measurement device identifier is a unique name, which refers to both the signal a record belongs to and the measurement device type. It connects the record in the primary processing data file to its related record in the raw measured data and measured data files.

- `measurement device data`
  These fields contain data that characterizes the measurement device, for example its type, maufacturer, measurement frequency, measurement range, bit length of its raw measured value, its status information etc.

- `scaling factors`
  The constant parameters needed to compute a scaled measured value form the raw measured value sent by the measurement device are stored here along with the type of the formulae used for scaling.

- `limits` (safety, warning)
  Soft safety and warning limits (both upper and lower ones) are given here, if they exist. These data are needed to perform limit checking in the primary/secondary data processing function of a computer controlled system.

- `filtering constants and processing characteristics`
  Constant parameters and formula/algorithm identifiers are given here for the following primary/secondary data processing and process monitoring functions:

  - filtering
  - averaging
  - computation of process trends

### 5.1.3 EVENTS DATA FILE

Events are stored in a finite length (measured in number of records) data file with a circular read and write pointer allow for incrementally increasing number of events to be received. A special event archiving method stores the older events in a correct time order.

The following fields are present in the records of this file.

- `time stamp`
  Shows the time when the event message was generated.

- `event type`
  This is a unique identifier of the event category (such as warning limit exceeded, equipment off-line, operator intervention etc.) the particular event belongs to.

- `sender`
  The identifier of the task in the computer control system that has generated the particular event message.

- `measurement device identifier(s)`
  Measurement device identifier(s) related to a particular event are given here. In case of a "warning limit exceeded" event, for example, we have the measurement device identifier of the signal the value of which has exceeded that particular warning limit.

- `other event specific data`
  In the case of the example above, here we have the warning limit value that has been exceeded.

### 5.1.4 ACTUATOR DATA FILE

The actuator data file is an "output data file" of a computer controlled system in the sense that it contains the value of actuators set by the controller tasks.

In some applications, however, not every actuator is equipped with a built-in sensor to provide us with feedback on the actual position of the sensor device. If such a built-in sensor exists, it is handled as an independent measurement device administered by the measurement device handling task. Its raw measured data record is then put into the raw measured data file and only a reference is made in the actuator data file.

A record in the actuator data file contains the following fields.

- `actuator device identifier`
  The actuator device identifier is a unique name, which refers to both the signal this record belongs to and to the actuator device type.

- `actuator position (set value)`
  The actuator position is a raw data value computed by controllers

on the basis of the properties of the actuator device. It is unscaled, i.e. raw data, which can be directly transferred to the actuator in question.

- `related measurement device identifier`
  If a built-in sensor is available to signal the actual position of the actuator (which may be different from its set value) then the measurement device identifier of this sensor is put here. It connects this actuator data record to a related record in the raw measured data, measured data and primary processing data files.

- `time stamp`
  This field shows the time when the set value command was issued to the actuator.

## 5.2 TYPICAL TASKS OF COMPUTER CONTROLLED SYSTEMS

Besides standard tasks like the scheduler and the real-time database manager, a computer controlled system that is a real-time software system contains the following special tasks.

### 5.2.1 MEASUREMENT DEVICE HANDLING

This task receives data from sensors, administers the states of sensors and puts received data into the *raw measured data* file. Most sensors are intelligent in the sense that they

- do not require regular data queries or acquisition but they only send data and cause a real-time interrupt, when signal changes occur,
- sense their status and send information on this self-diagnosis in the status attached to every measured value.

### 5.2.2 PRIMARY AND SECONDARY PROCESSING

This task performs primary and secondary data processing including scaling, handling missing or invalid data, limit checking, filtering, averaging etc. These functions are described in section 3.. of this chapter.

Process monitoring functions, such as logsheet generation, computation of process trends and alarm generation belong to this task, too.

### 5.2.3 EVENT HANDLING

Besides process or plant events and operator actions which are signalled by the *measurement device handling*, *primary processing*, *secondary processing* or *controller* tasks, every software error generates an event message. These messages are sent to the *event handling* tasks via a one-way *send* communication primitive.

The event handling task handles and administers received event messages, puts them into the *event* circular file in the correct time order and takes care of their archiving. It also supports the logsheet and alarm report functions in retrieving events of prescribed types, over any desired time interval or according to other user defined filtering viewpoint.

### 5.2.4 CONTROLLER(S) AND ACTUATOR HANDLING

Controllers implement the active control tasks defined in the computer controlled system in question. They use measured data to compute actuator data to be sent to the controlled system according to their control algorithm.

The actuator handling task administers the state of system actuators and downloads their required position to the actuator devices. It also senses actuator status and notifies the software system and controllers via events in the case of any failure or fault.

# Appendix B
# THE COFFEE MACHINE

The tools and techniques introduced in various Chapters of the book are explained and demonstrated using the same simple example which is the subject of this Appendix. This way it is possible to compare different, sometimes alternative or competitive methods.

This common example, a coffee machine seen in Fig. B.1, is one of the simplest process systems to be controlled from the system modelling point of view, yet is well-known in everyday life.

The required dynamic state-space model equations for the coffee machine are developed in two main steps.

1. *Specification of the modelling task*
   includes the specification and modelling goal(s) of the coffee machine as a dynamic system.

2. *Development of model equations*
   using first engineering principles.

For more about systematic modelling methodology, see [96].

## 1. SYSTEM DESCRIPTION

The description of a system to be modelled is prepared in the following way. First, we specify system boundaries, which separate the system from its environment and describe the processes and interactions considered within the system and between the system and its environment. Then the input and output signals are described, together with the operating region of interest. We usually put the main elements of a system description on a so called *flowsheet*, which is a schematic picture of the system to be modelled with its boundaries, main sub-systems and signals.

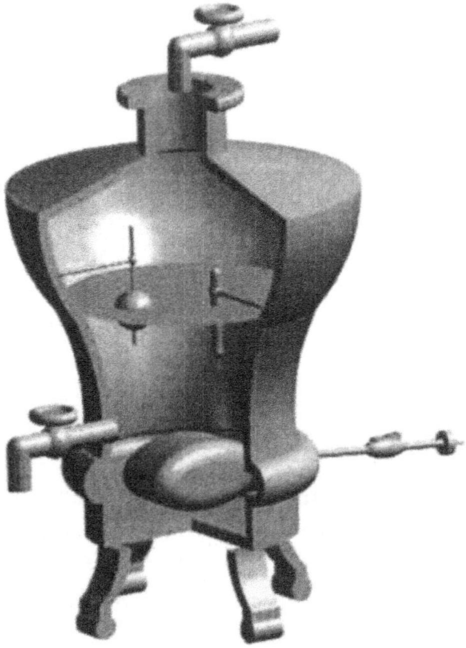

*Figure B.1.* Coffee machine

The *modelling goal* which, influences the precision and the type of the model to be used is also usually briefly described.

**System description for the coffee machine**
Consider a perfectly stirred tank with water flowing in and out. The in- and outflow is controlled by valves. Let us assume that the tank is adiabatic, i.e. its walls are perfectly insulated and moreover it also contains an electric heater, which is controlled by a switch. The flowsheet is shown in Figure B.2.

**Modelling goal**
We want to have a model of the coffee machine for diagnosis and control. This implies that a dynamic model with moderate complexity and precision is needed to describe the dominant time constants of the system.

In particular we want to examine different operating procedures, that is sequential and perhaps parallel operator actions, which lead to optimal coffee making in terms of time and energy, for example.

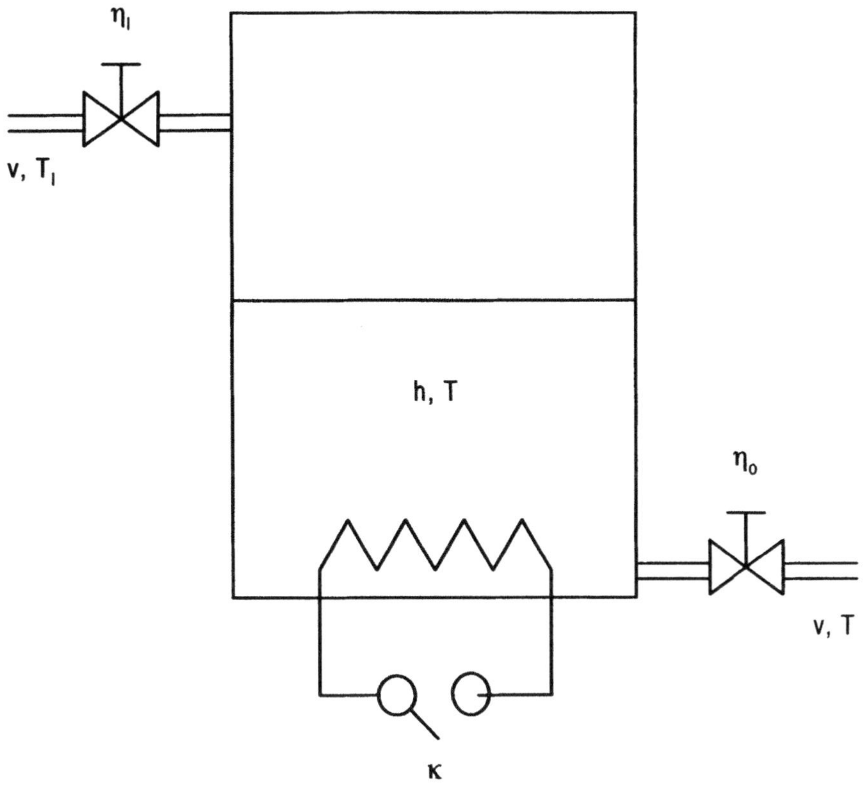


*Figure B.2.* The flowsheet of the coffee machine

**Operating region**

The above modelling goal implies that we only consider such system states when we have water in the coffee machine, that is when it is not empty or not overheated containing only vapour.

# 2. DYNAMIC MODEL EQUATIONS

The dynamic model equations of the coffee machine are derived from conservation balance equations for the overall mass and the energy of the system equipped by suitable algebraic constitutive equations. In order to have a relatively simple dynamic model suitable for control and diagnostic purposes, simplification assumptions are needed. These are the following.

**Modelling assumptions**

1. The liquid in the tank is perfectly stirred.

2. There is only water in the tank.
3. Balances are only set up for the liquid phase (the gas phase is neglected).
4. Physico-chemical properties are constant.
5. There are binary valves and switches.
6. The tank is cylindrical with a constant cross-section $A$.
7. The properties of water at the outlet are the same as those of the water in the tank.
8. The tank walls are perfectly insulated (adiabatic tank).

## 2.1 DIFFERENTIAL (BALANCE) EQUATIONS

*Conservation balance for the overall mass*

$$\frac{dh}{dt} = \frac{v}{A}\eta_I - \frac{v}{A}\eta_O \tag{B.1}$$

*Conservation balance for the energy*

$$\frac{dT}{dt} = \frac{v}{Ah}(T_I - T)\eta_I + \frac{H}{c_p \rho h}\kappa \tag{B.2}$$

where the variables are

| | |
|---|---|
| $t$ | time $[s]$ |
| $h$ | level in the tank $[m]$ |
| $v$ | volumetric flowrate $[m^3/s]$ |
| $c_p$ | specific heat $[Joule/kgK]$ |
| $\rho$ | density $[kg/m^3]$ |
| $T$ | temperature in the tank $[K]$ |
| $T_I$ | inlet temperature $[K]$ |
| $H$ | heat provided by the heater $[Joule/sec]$ |
| $A$ | cross section of the tank $[m^2]$ |
| $\eta_I$ | binary input valve $[1/0]$ |
| $\eta_O$ | binary output valve $[1/0]$ |
| $\kappa$ | binary switch $[1/0]$ |

**Initial conditions**

$$h(0) = h_0 \quad , \quad T(0) = T_0$$

**Mathematical properties**
The model equations above form a set of nonlinear ordinary differential equations with suitable initial conditions.

## 2.2 SYSTEM VARIABLES

The conservation balance equations of any process system determine its state equations, therefore Eqs. (B.1)-(B.2) can be seen as the state equations of the coffee machine.

From a system theoretical point of view, the above model equations form a nonlinear concentrated parameter time-invariant state-space model of a process system with two state variables

$$x = [\, h \, , \, T \,]^T$$

and three potential input variables

$$u_P = [\, \eta_I \, , \, \eta_O \, , \, \kappa \,]^T$$

The potential input variables influence the behaviour of the coffee machine, but the actual measurement and actuator devices determine whether they will be actuator or disturbance variable.

The process instrumentation diagram, which is an extension to the process flowsheet, contains the measurement devices and actuators available in the processing unit. From this we can determine which variables will contain the set of output, actuator and disturbance variables.

An output variable can be any variable which is directly measurable and contains information about the state variables of the process. In the case of the coffee machine, we may assume that we have both level and temperature sensors to measure both of the state variables. This way the output variable vector is as follows:

$$y = x = [\, h \, , \, T \,]^T$$

A potential input variable can be an actuator variable if we have a real actuator (a switch, a motor, a valve etc.) to set its value as desired. In the case of the coffee machine we have already assumed that we have binary switches to set all of the three potential input variables, therefore the actuator variables will be:

$$u = u_P = [\, \eta_I \, , \, \eta_O \, , \, \kappa \,]^T$$

In real life, however, not every actuator is equipped with a built-in sensor to provide us with feedback on the actual position of the sensor device. If such a measured value about the position of the actuator is not available, we need to use diagnostic methods to infer on the status of the actuator device.

The built-in sensor, if available, is treated as an independent sensor.

# References

[1] Gupta, M. M., Sinha, N. S. (Eds.) (1996) *Intelligent Control Systems. Theory and Applications*, IEEE Press, New York.

[2] Antsaklis, P. J., Passino, K. M. (Eds.) (1993) *An Introduction to Intelligent and Autonomous Control*, Kluwer Academic Publishers, Norwell, MA.

[3] Antsaklis, P. J. (1994) Defining Intelligent Control. *IEEE Control Systems Magazine*, **14**(3), pp. 4-66.

[4] Ginsberg, M. (1993) *Essentials of Artificial Intelligence*, Morgan Kaufmann Pub.

[5] Russell, S., Norvig, P. (1995) *Artificial Intelligence - A Modern Approach* In: *Series in Artificial Intelligence*, Prentice-Hall International, Inc.

[6] Poole, D., Mackworth, A., Goebel, R. (1998) *Computational Intelligence - A Logical Approach*, Oxford University Press.

[7] Nilsson, N. J. (1980) *Principles of Artificial Intelligence*, Morgan Kaufmann Pub.

[8] Winston, P. H. (1992) *Artificial Intelligence* (3rd edition), Addison-Wesley Pub. Co.

[9] Stephanopoulos, G., Han, C. (1996) Intelligent Systems in Process Engineering. *Computers and Chemical Engineering*, **20**(6-7), pp. 743-791.

[10] Linkens, D. A., Chen, M. Y. (1995) Expert Control Systems 1. Concepts, Characteristics and Issues. *Engineering Applications of Artificial Intelligence*, **8**, pp. 413-421.

[11] Buchanan, B., Shortlife, E. H. (1984) *Rule-Based Expert Systems, MYCIN*, Addison-Wesley, Reading, MA.

[12] Zeigler, B. P. (1987) Knowledge Representation from Newton to Minsky and beyond. *Applied Artificial Intelligence*, **1**, pp. 87-107.

[13] Sowa, J. F. (1999) *Knowledge Representation: Logical, Philosophical, and Computational Foundations*, Pws Pub. Co.

[14] Ullman, J. D. (1988) *Principles of Database and Knowledge-Base Systems*, Computer Science Press.

[15] Meyer, B. (2000) *Object-Oriented Software Construction, 2nd Edition*, Prentice Hall.

[16] Minsky, M. A. (1975) Framework for Representating Knowledge. In: *The Psychology of Computer Vision* (Ed: Winston, P.), McGraw-Hill, New York, pp. 211-277.

[17] Hayes, P. J. (1980) The Logic of Frames. In: *Frame Conceptions and Text Understanding* (Ed: Metzing, D., Walter de Gruyter), Berlin, pp. 46-61.

[18] Fagin, R. Y., Halpern, J., Moses, Y., Vardi, M. Y. (1994) *Reasoning about Knowledge*, MIT Press.

[19] Kolodner, J. (1993) *Case-Based Reasoning*, Morgan Kaufmann Pub.

[20] Barr, A., Feigenbaum, E. (1981) *The Handbook of Artificial Intelligence, Volume I.*, Morgan Kaufmann Pub.

[21] Renesereth, M. R., Nilsson, N. J. (1987) *Logical Foundation of Artificial Intelligence*, Morgan Kaufmann Pub.

[22] Lunardhi, A.D., Passino, K. M. (1991) Verification of Dynamic Properties of Rule-based Expert Systems, *Proc. of IEEE Conf. on Decision and Control*, (Brighton, UK.), pp. 1561-1566.

[23] Gupta, U. (Ed.) (1991) *Validating and Verifying Knowledge-Based systems*, IEEE Computer Society Press, Loas Alamitos, CA.

[24] Greissman, J. R. (1988) Verification and Validation of Expert Systems. *AI Expert*, **3**, pp. 26-33.

[25] Perkins, V. A., Laffey, T. J., Pecora, D., Nguyen, T. A. (1989) Knowledge Base Verification. In: *Topics in Expert System Design* (Eds: Guida, G., Tasso, C.), Elsevier, North Holland.

[26] Passino, K. M., Lunardhi, A.D. (1995) Qualitative Analysis of Expert Control Systems. In: *Intelligent Control: Theory and Applications* (Eds: Gupta, N. N., Sinha, N. K.), IEEE Press, New York, pp. 404-442.

[27] Suwa, M., Scott, A.C., Shortliffe, E. H. (1982) An Approach to Verifying Completeness and Consistency in a Rule-Based Expert System. *AI Magasine*, pp. 16-21.

[28] Leeuwen, van J. (1990) *Handbook of Theoretical Computer Science, Vol. A., Algorithms and Complexity*, Elsevier - MIT Press, Amsterdam.

[29] Kim, S. (1988) Checking a Rule Base with Certainty Factor for Incompleteness and Inconsistency. In: *Uncertainty and Intelligent Systems, Lecture Notes in Computer Science No. 313*, Springer-Verlag, New York.

[30] Patridge, D. (1987) The Scope and Limitations of First Generation Expert Systems. *Future Generation Computer System*, North Holland, Amsterdam, pp. 1-10.

[31] Guy, L., Steele, Jr. (1990) *Common Lisp: The Language* (2nd edition), Digital Press.

[32] Winston, P. H., Horn, B. K. P. (1993) *Lisp* (3rd edition), Addison-Wesley Pub Co.

[33] Graham, P. (1995) *ANSI Common Lisp*. In: *Series in Artificial Intelligence*, Prentice-Hall International, Inc.

[34] Tanimoto, S. L. (1995) *The Elements of Artificial Intelligence Using Common Lisp* (2nd edition), W. H. Freemann & Co.

[35] Queinnec, C., Callaway, K. (1996) *Lisp in Small Pieces*, Cambridge University Press.

[36] Sterling, L., Shapiro, E. (1994) *The Art of Prolog: Advanced Programming Techniques*. In: *MIT Press Series in Logic Programming*, MIT Press.

[37] Bratko, I. (1990) *Prolog Programming for Artificial Intelligence* (2nd edition), Addison-Wesley Pub Co.

[38] Clocksin, W. F., Mellish, C. S. (1994) *Programming in Prolog*, Springer-Verlag.

[39] Civington, M. A., Nute, D., Vellino, A. (1996) *Prolog Programming in Depth*, Prentice-Hall.

[40] Van Le, T. (1993) *Techniques of Prolog Programming with Implementation of Logical Negation and Quantified Goals*, John Wiley & Sons, Inc.

[41] Ratledge, E. C., Jacoby, J. E. (1990) *Handbook on Artificial Intelligence and Expert Systems in Law Enforcement*, Greenwood Publishing Group.

[42] Ignizio, J. P. (1991) *An Introduction to Expert Systems*, McGraw-Hill Higher Education.

[43] Jackson, P. (1999) *Introduction to Expert Systems* (3rd edition), In: *International Computer Science Series*, Addison-Wesley Pub. Co.

[44] Giarrarano, J. C. (1998) *Expert Systems: Principles and Programming* (3rd edition), Pws Pub. Co.

[45] Durkin, J. (1998) *Expert Systems: Design and Development*, Prentice Hall.

[46] Musliner, D. J., Hendler, J. A., Agrawala, A. K., Durfee, E. H., Strosnider, J. K., Paul, C. J. (1995) The Challenges of Real-Time AI. *Computer*, **28**, pp. 58-66.

[47] Aström, K. J., Anton, J., Arzén, K. E. (1986) Expert Control. *Automatica*, **22**, pp. 277-286.

[48] Rodd, M. G., Holt, J., Jones, A. V. (1993) Architectures for Real-Time Intelligent Control Systems. *IFIP Transactions B - Applications in Technology*, **14**, pp. 375-388.

[49] Williams, J. G., Jouse, W. C. (1993) Intelligent Control in Safety Systems. *IEEE Transactions on Nuclear Science*, **40**, pp. 2040-2044.

[50] Pang, G. K. H. (1991) A Framework for Intelligent Control. *Journal of Intelligent and Robotic Systems*, **4**(2), pp. 109-127.

[51] Abbod, M. F., Linkens, D. A., Browne, A. Cade, N. (2000) A Blackboard Software Architecture for Integrated Control Systems. *Kybernetes*, **29**, pp. 999-1015.

[52] Linkens, D. A., Abbod, M. F., Browne, A. Cade, N. (2000) Intelligent Control of a Cryogenic Cooling Plant based on Blackboard System Architecture. *ISA Transactions*, **39**, pp. 327-343.

[53] Bergin, T., Khosrowpour, M., Travers, J. (1993) *Computer-Aided Software Engineering : Issues and Trends for the 1990s and Beyond*, Idea Group Publishing.

[54] Weiss, S. M., Kulikowski, C. A. (1984) *A Principal Guide to Designing Expert Systems*, Rowmann and Allenheld, NJ.

[55] Payne, E. C., McArthur, R. (1990) *Developing Expert Systems : A Knowledge Engineer's Handbook for Rules and Objects*, John Wiley and Sons.

[56] Hangos, K. M. (1991) Qualitative Process Modelling. In: *Chemical Process Control, CPCIV* (Eds. Arkun, Y., Ray, W. H.), AICHE CACHE, pp. 209-236.

[57] Feraybeaumont, S., Corea, R., Tham, M. T., Morris, A. J. (1992) Process Modelling for Intelligent Control. *Engineering Applications of Artificial Intelligence*, **5**, pp. 483-492.

[58] Faltings, B., Struss, P. (1992) *Recent Advances in Qualitative Physics*, The MIT Press, Cambrdge, MA.

[59] Weld, D. S., de Kleer, J. (Eds.) (1990) *Readings in Qualitative Reasoning about Physical Systems*, The Morgan Kaufman.

[60] Kuipers, B. (1986). Qualitative Simulation. *Artificial Intelligence*, **29**, pp. 289–388.

[61] Forbus, K.D. (1984), Qualitative Process Theory, *Artificial Intelligence*, **24**, pp. 85–168.

[62] Reinschke, K.J. (1988) *Multivariable Control. A Graph-theoretic Approach.* In: *Lecture Notes in Control and Information Sciences* (ed. M. Thoma and A. Wyner), Springer Verlag.

[63] Moore, R. E.. (1966) *Interval Analysis*, Prentice Hall Series in Automatic Computation.

[64] Nguyen, H. T., Kreinovich, V., Zuo, Q. (1997) Interval-Valued Degrees of Belief: Applications of Interval Computations to Expert Systems and Intelligent Control. *International Journal of Uncertainty Fuzzyness and Knowledge-Based Systems*, **5**, pp. 317-358.

[65] Kuipers, B. (1989) Qualitative Reasoning: Modeling and Simulation with Incomplete Knowledge. *Automatica*, **25**, pp. 571-585.

[66] Hangos, K. M., Csáki, Zs., Jorgensen, S. B. (1992) Qualitative Simulation in the Limit. *Artificial Intelligence in Engineering*, **7**(2), pp. 105-109.

[67] Hangos, K. M., Csáki, Zs. (1992) Qualitative Model-Based Intelligent Control of a Distillation Column. *Engineering Application of Artificial Intelligence*, **5**, pp. 431-440.

[68] Murota, K. (1987) *Systems Analysis by Graphs and Matroids*, Springer-Verlag, Berlin.

[69] Puccia, C.J., Levins, R. (1985) *Qualitative Modelling of Complex Systems: An Introduction to Loop Analysis and Time Averaging*, Harward University Press, Cambridge (Massachusetts) - London (England)

[70] Rose, P., Kramer, M. A. (1991) Qualitative Analysis of Causal Feedback, In: *Proc. 10th National Conference on Artificial Intelligence (AAAI-91)*, Anaheim, CA.

[71] Venkatasubramanian, V., Vaidhyanathan, R. (1994) A Knowledge-based Framework for Automating HAZOP Analysis. *AIChE Journal*, **40**, pp. 496-505.

[72] Gál, I.P., Hangos, K.M. (1998) SDG Model-based Structures for Fault Detection. In: *Preprints IFAC Workshop on On-line Fault Detection and Supervision in Chemical Process Industries*, Lyon (France), Vol. 1. p. 6.

[73] Petri, C. A. (1962) *Kommunikation mit Automaten*, Institute für Instrumentelle Mathematik, Schriften des IIM, Nr 3.

[74] Yamalidou, E. C., Kantor, J. C. (1991) Modeling and Optimal Control of Disctere-Event Chemical Process Systems. *Computers chem. Engng*, **15**(7), pp. 503-519.

[75] Pages, A., Pingaud, H. (1995) An Hybrid Process Model based on Petri Nets Applied to Short Term Scheduling of Batch/Semi Continuous Plants. *Workshop Analysis and Design of Event-Driven Operations in Process Systems*, Imperial College, London, UK.

[76] Gerzson, M., Hangos, K. M. (1995) Analysis of Controlled Technological Systems using High Level Petri Nets. *Comp. Chem. Engng*, **19**(Suppl), pp. S531-S536.

[77] Moody, J. O., Antsaklis, P. J. (1998) *Supervisory Control of Discrete Event Systems Using Petri Nets*, Kluwer International Series on Discrete Event Dynamic Systems, 8.

[78] Murota, T. (1989) Petri Nets: Properties, Analysis and Applications. *Proceedings of the IEEE*, **77**(4), pp. 541-580.

[79] Peterson, J. L. (1981) *Petri Net Theory and the Modeling of Systems*, Prentice-Hall.

[80] Wang, J. (1998) *Timed Petri Nets: Theory and Application*, Kluwer International Series on Discrete Event Dynamic Systems, 9.

[81] Cox, E. (1994) *The Fuzzy Systems Handbook*, AP Professional, Boston.

[82] Jantzen, J. (1994) *Fuzzy Control*, Lecture Notes in On-Line Process Control, Technical University of Denmark, Lyngby, Denmark.

[83] Zadeh, L. (1965) Fuzzy sets. *Inf. and Control*, **8**, pp. 338-353.

[84] Zimmermann, H.-J. (1993) *The Fuzzy Set Theory and Its Applications*, Kluwer, Boston.

[85] Mamdani E. H. (1977) Application of fuzzy logic to approximate resoning. *IEEE Trans. Computers*, **26**(12), pp. 1182-1191.

[86] Lee C. C. (1990) Fuzzy Logic in Control Systems: Fuzzy Logic Controller. *IEEE Trans. on Systems, Man and Cybernetic*, **20**(2), pp. 404-435.

[87] Taur, J. (1999) Design of Fuzzy Controllers with Adaptive Rule Insertion. *IEEE Trans. on Systems, Man and Cybernetic Part B Cybernetics*, **29**(3), pp. 389-397.

[88] Wang, W. J., Tang, B. Y. (1999) A Fuzzy Adaptive Method for Intelligent Control. *Expert Syst. Appl*, **16**(1), pp. 43-48.

[89] Pedrycz, W. (1993) *The Fuzzy Control and Fuzzy Systems*, Wiley and Sons.

[90] G2 Reference Manual (for G2 Version 3.0) (1992) Gensym Corporation.

[91] http://www.gensym.com/manufacturing/g2-overview.shtml

[92] Aström, K. J., Wittenmark, B. (1990) *Computer Controlled Systems*, Prentice Hall, New York, London, Toronto, Sydney, Tokyo, Singapore.

[93] Hangos, K. M., Bokor, J., Gerzson, M. (1995) *Computer Controlled Systems*, Veszprém University Press, Veszprém.

[94] Kailath, K. (1980) *Linear Systems*, Prentice Hall, New York, London, Toronto, Sydney, Tokyo, Singapore.

[95] Braek, R., Haugen, O. (1994) *Engineering Real Time Systems*, Prentice Hall, New York, London, Toronto, Sydney, Tokyo, Singapore.

[96] Hangos, K. M., Cameron, I. T. (2001) *Process Modelling and Model Analysis*, Academic Press, New York.

# Index


$\pi$-curve, *see* fuzzy system, membership function, $\pi$-curve

A* search, *see* search, A*
action, 32, 35, 36
  of a rule, *see* rule, consequence
actuator data file, 260, 271
actuator handling, 273
AI, *see* artificial intelligence
AI models, *see* qualitative models
algorithmic complexity
  nondeterministic polynomial, *see* algorithmic complexity, NP-complete
  NP-complete, 15, 38, 46, 61, 64–66, 68, 141, 186
  polynomial, 61, 64, 68, 131
analysis
  of Petri nets, *see* Petri nets, analysis
  of rule-base, 59–68
    completeness, 60, 64–66, 147, 220
    contradiction freeness, 60–64, 66, 147
    of fuzzy system, *see* fuzzy system, rule-base analysis
AND-OR graph, 44, 91
  hyperarc, 44, 91
  hyperpath, 45
approximation, *see* fuzzy system, linguistic modifier, approximation
arbitrary curve, *see* fuzzy system, membership function, arbitrary curve
arc
  of AND-OR graph, *see* AND-OR graph, hyperarc
  of Petri nets, *see* Petri nets, arc
  of search graph, 36
artificial intelligence, 1, 2, 15, 85, 127, 145
atom
  in Lisp, *see* Lisp, atom
  in Prolog, *see* Prolog, atom
attribute, 23
  in G2, *see* G2, attribute
attribute table (in G2), *see* G2, attribute table

backtracking, 40, 43, 52
  in Prolog, *see* Prolog, backtracking
backward chaining, *see* reasoning, backward
backward reasoning, *see* reasoning, backward
  in G2, *see* G2, reasoning, backward
  in Prolog, *see* Prolog, reasoning
behaviour tree, *see* qualitative simulation, behaviour tree
bell-shaped curve, *see* fuzzy system, membership function, bell-shaped curve
bidirectional reasoning, *see* reasoning, bidirectional
blind search, *see* search, blind
body (of a rule), *see* Prolog, rule, body
boundedness (of Petri nets), *see* Petri nets, properties, boundedness
branch, 36, 40, 186
breadth-first search, *see* search, breadth-first

casual system, *see* system, casual system
class, 23
  in G2, *see* G2, class
clause (in Prolog), *see* Prolog, clause
CNF, *see* conjunctive normal form
COA (centre of area), *see* fuzzy system, defuzzification methods, COA
coffee machine, 275–279
  behaviour tree, 142
  class hierarchy, 25
  confluences, 146

dynamic model equations, 277
flowsheet, 275
in G2, 228
QDE model, 137, 139
semantic nets, 29
Signed Directed Graph (SDG) model, 150
state-space model, 150
system description, 275
system variables, 279
complement (of fuzzy sets), *see* fuzzy system, fuzzy set operation, complement
completeness, *see* analysis, of rule-base, completeness
composition (of fuzzy sets), *see* fuzzy system, fuzzy set operation, composition
computer-controlled system, 1, 2, 153
concentrated parameter system, *see* system, concentrated parameter system
concurrency, *see* Petri nets, parallelism, concurrency
condition
in Petri nets, *see* Petri nets, place, input
of a rule, *see* rule, condition
conflict, 36
in Petri nets, *see* Petri nets, parallelism, conflict
conflict resolution, 36, 51
confluences, *see* qualitative models, confluences
confusion, *see* Petri nets, parallelism, confusion
conjunctive normal form, 17
connection, 32
in G2, *see* G2, connection
consequence
in Petri nets, *see* Petri nets, place, output
of a rule, *see* rule, consequence
conservation (of Petri nets), *see* Petri nets, properties, conservation
conservative Petri nets, *see* Petri nets, conservative Petri nets
consistency, *see* fuzzy system, rule-base analysis, consistency
continuous time system, *see* system, continuous time system
contradiction freeness, *see* analysis, of rule-base, contradiction freeness
controllability, *see* dynamical properties, controllability of LTI systems
controllers, 192, 273
in fuzzy system, *see* fuzzy system, fuzzy controller
controls (in G2), *see* G2, controls
counterfactual reasoning, *see* reasoning, counterfactual
coverability (of Petri nets), *see* Petri nets, properties, coverability
crisp set, 192, 194

data server (in G2), *see* G2, data server
data source (in G2), *see* G2, data server
data-driven chaining, *see* reasoning, forward
database, 3, 11–15
in Lisp, *see* Lisp, database
in Prolog, *see* Prolog, database
relational, 12, 14, 85
database manager, 3, 14
datalog rule, *see* rule, datalog
declarative programming language, *see* Prolog
decomposition, 67–68
heuristic, 68
hierarchical, 67
of Petri nets, *see* Petri nets, decomposition
strict, 68
defuzzification, *see* fuzzy system, defuzzification
defuzzifier, *see* fuzzy system, fuzzy controller, defuzzifier
dependence graph
of datalog rule set, 21–22, 61, 64, 68
strong components, 68
depth-first search, *see* search, depth-first
design (of fuzzy controller), *see* fuzzy system, design (of fuzzy controller)
developers' interface, 5, 31
in expert system, *see* expert system, developers' interface
in G2, *see* G2, developers' interface
diagnosis, 47
dilution, *see* fuzzy system, linguistic modifier, dilution
discrete time system, *see* system, discrete time system
disjunctive normal form, 17
displays (in G2), *see* G2, displays
DNF, *see* disjunctive normal form
dynamical properties
asymptotic stability of LTI systems, 257
BIBO stability of LTI systems, 257
controllability of LTI systems, 180, 255
observability of LTI systems, 256
realization property, 255

stability of LTI systems, 257

enabled (transition), *see* Petri nets, transition, enabled
end-user controls (in G2), *see* G2, controls
end-user interface (in G2), *see* G2, end-user interface
events data file, 270
execution (of a rule), *see* rule, execution
expansion, 53
expert system, 5, 31, 85, 103, 109
  case specific database, 104
  developers' interface, 105
  explanation subsystem, 105
  G2, *see* G2
  inference engine, 104
  knowledge acquisition subsystem, 105
  knowledge base, 104
  real-time expert system, 47, 109–126, 227–250
  user interface, 105
expert system shell, 103–107
  G2, *see* G2
explanation subsystem, *see* expert system, explanation subsystem
explanative reasoning, *see* reasoning, explanative
expression, 18
  in Lisp, *see* Lisp, expression
external interface (in G2), *see* G2, external interface

fact, 12, 32–34, 36, 38, 44, 45
  in Prolog, *see* Prolog, fact
file, 13, 15
finite capacity Petri nets, *see* Petri nets, finite capacity Petri nets
firing
  of a rule, *see* rule, firing
  of a transition, *see* Petri nets, transition, firing
flowsheet, 275
FOM (first of maxima), *see* fuzzy system, defuzzification methods, FOM
formula
  in G2, *see* G2, formula
  simulation formula (in G2), *see* G2, simulation formula
forward chaining, *see* reasoning, forward
forward reasoning, *see* reasoning, forward
  in G2, *see* G2, reasoning, forward
frame, 12, 26–27
function
  in G2, *see* G2, function
  in Lisp, *see* Lisp, procedure
  of Petri nets, *see* Petri nets, function

fuzzification, *see* fuzzy system, fuzzification
fuzzy controller, *see* fuzzy system, fuzzy controller
fuzzy set, *see* fuzzy system, fuzzy set
fuzzy system, 191–226
  defuzzification, 225
  defuzzification methods
    COA (centre of area), 225
    FOM (first of maxima), 226
    height, 226
    LOM (last of maxima), 226
    MOM (mean of maxima), 225
  design (of fuzzy controller), 216
  fuzzification, 223
  fuzzy controller, 215–226
    defuzzifier, 216
    inference engine, 223
    postprocessing, 216, 225
    preprocessing, 216, 223
    rule-base, 216
  fuzzy controller types
    neuro-fuzzy, 216, 220
    PID, 216, 219
    self-organizing, 216, 220
    table based, 216
  fuzzy set, 192–215
    discrete representation, 198
    inference, 208, 214
    normalized, 196
    relation, 209
    support, 196
  fuzzy set operation, 200–208
    complement, 200
    composition, 211, 214
    implication, 211
    intersection, 200
    union, 200
  grade of membership, 194
  linguistic modifier, 206–208
    approximation, 206
    dilution, 206
    intensification, 206
    restriction, 206
  linguistic variable, 205
  max-min composition, 211
  membership function, 194, 217
    $\pi$-curve, 197
    arbitrary curve, 198
    bell-shaped curve, 196, 206
    irregularly shaped curve, 198
    linear representation, 197
    s-curve, 197, 203
    shouldered curve, 198
    triangular shape curve, 197, 218
    z-curve, 197, 203
  outer product, 222

rule, 208, 214
rule-base
normalized, 219
standard, 219
rule-base analysis, 220–223
completeness, 220
consistency, 221
interaction, 222
redundancy, 222
singleton, 199, 218, 225, 226

G2, 227–250
attribute, 228, 233, 236, 242
attribute table, 228, 231, 236, 237, 242
class, 228, 230–231, 241
connection, 229–231, 234–235
controls, 230, 247–249
action button, 248
check box, 248
radio button, 248
slider, 249
type-in box, 249
data server, 229, 230, 233, 240
data source, *see* G2, data server
developers' interface, 241–247
access control, 247
debugging, 246
describing, 245
icon editor, *see* G2, icon editor
inspecting, 244
text editor, *see* G2,text editor
tracing, 246
displays, 230, 243, 247–248
chart, 247
dial, 247
free-form table, 248
meter, 247
readout table, 247
end-user controls, *see* G2, controls
end-user interface, 247–249
controls, *see* G2, controls
displays, *see* G2, displays
message, *see* G2, message
external interface, 250
formula, 229, 234
simulation, *see* G2, simulation formula
function, 231, 238–239
hierarchy, 231, 232
icon, 231, 241, 243–244
icon editor, 241, 243–244
inference engine, 229, 234, 239–240
item, 230
knowledge base, 230–239, 244–245
logbook, 230, 247, 249
message, 230, 247, 249
message board, 230, 247, 249
object, 228–231, 241
parameter, 229, 230, 233–234
procedure, 229, 231, 237–238
reasoning, 227, 239–240
backward, 229, 234, 239
forward, 229, 240
relation, 230, 234–235
rule, 229, 230, 235–236
generic, 231, 235, 236
simulation formula, 229, 240
simulator, 227, 229, 234, 239–241
text editor, 241–242
variable, 229, 230, 233–234, 240
workspace, 228, 230, 232
generalized modus ponens, *see* modus ponens, generalized
generic rule (in G2), *see* G2, rule, generic
goal, 33, 34, 38, 40, 45, 51, 52, 55
in Prolog, *see* Prolog, goal
goal-driven chaining, *see* reasoning, backward
grade of membership, *see* fuzzy system, grade of membership
graph
AND-OR, *see* AND-OR graph
of Petri nets, *see* Petri nets, graph representation
search, *see* search graph
Signed Directed Graph (SDG), *see* Signed Directed Graph (SDG) models

head (of a list)
in Lisp, *see* Lisp, list, head
in Prolog, *see* Prolog, list, head
head (of a rule), *see* Prolog, rule, head
height, *see* fuzzy system, defuzzification methods, height
heuristic, 3, 15, 32, 37, 51, 52, 55, 56
heuristic search, *see* search, heuristic
heuristic knowledge, 104
hierarchical system, *see* system, hierarchical system
hierarchy, 23, 25, 26
in G2, *see* G2, hierarchy
hill-climbing search, *see* search, hill-climbing
Horn clause, 88
hyperarc, *see* AND-OR graph, hyperarc
hypergraph, *see* AND-OR graph
hyperpath, *see* AND-OR graph, hyperpath
hypothetical reasoning, *see* reasoning, hypothetical

icon (in G2), *see* G2, icon
icon editor (in G2), *see* G2, icon editor

identification, 47
implication, 16, 19, 82, 88, 99
  of fuzzy sets, *see* fuzzy system, fuzzy set operation, implication
implicative normal form, 17
incidence matrix (of Petri nets), *see* Petri nets, matrix representation
INF, *see* implicative normal form
inference (on fuzzy set), *see* fuzzy system, fuzzy set, inference
inference engine, 4, 31, 37, 51, 69, 113
  in expert system, *see* expert system, inference engine
  in fuzzy system, *see* fuzzy system, fuzzy controller, inference engine
  in G2, *see* G2, inference engine
  in Prolog, *see* Prolog, inference engine
infinite capacity Petri nets, *see* Petri nets, infinite capacity Petri nets
initial state, 32, 34, 38, 39, 45, 51, 52, 55
intelligent control, 2, 11
intensification, *see* fuzzy system, linguistic modifier, intensification
interaction, *see* fuzzy system, rule-base analysis, interaction
intersection (of fuzzy sets), *see* fuzzy system, fuzzy set operation, intersection
interval algebra, 130
interval calculus, 128
interval operation, *see* operation, interval
interval valued variable, *see* qualitative variable, interval valued
irregularly shaped curve, *see* fuzzy system, membership function, irregularly shaped curve
item (in G2), *see* G2, item

knowledge acquisition subsystem, *see* expert system, knowledge acquisition subsystem
knowledge base, 4, 11, 31, 113
  in G2, *see* G2, knowledge base
  of expert system, *see* expert system, knowledge base
  rule-based, 103
knowledge base maintenance, 31
knowledge base manager, 5, 113
knowledge engineer, 5, 106
knowledge extraction, 107
knowledge representation, 11 29
  expert system shells, *see* expert system shells
  G2, *see* G2
  Lisp, *see* Lisp
  Prolog, *see* Prolog
knowledge-based systems, 4

landmark set, *see* qualitative simulation, landmark set
linear system, *see* system, linear system
linguistic modifier, *see* fuzzy system, linguistic modifier
linguistic variable, *see* fuzzy system, linguistic variable
Lisp, 70–84
  atom, 70, 72
  built-in procedure, *see* Lisp, primitive
  database, 70
  evaluation, 72–73
  expression, 70
  list, 70, 72
    empty, 70, 74, 77–79
    head, 70
    tail, 70, 77
  NIL, *see* Lisp, list, empty
  number, 70
  predicate, *see* Lisp, primitive, predicate
  primitive, 72–82
    ', 73
    =, 77, 81
    >, 81
    **AND**, 79, 82
    **APPEND**, 75
    **ATOM**, 78
    **CAR**, 74
    **CDR**, 74
    **COND**, 80, 81, 83, 84
    **CONS**, 75, 83
    **DEFUN**, 81–83
    **ENDP**, 78
    **EQUAL**, 77
    **EQ**, 77
    **EVAL**, 83
    **FIRST**, 74, 75, 77, 78, 83, 84
    **IF**, 79
    **LENGTH**, 81
    **LISTP**, 78
    **LIST**, 75, 77
    **MEMBER**, 77, 79, 80
    **NOT**, 79–82
    **NULL**, 78–80, 83, 84
    **NUMBERP**, 78–81
    **OR**, 79, 82
    **QUOTE**, 73
    **REST**, 74, 75, 78, 83, 84
    **SETF**, 73, 76, 79, 80
    **SET**, 76
    **SYMBOLP**, 78
    **UNLESS**, 80
    **WHEN**, 80

arithmetic, 76–77
assignment, 76
conditional, 79–81
list-management, 74–76
predicate, 77–79
procedure definition, 81–82
procedure, 70–72
built-in, *see* Lisp, primitive
user-defined, 72, 81
program, 70
recursion, 83, 84
symbol, 70, 73
variable, 73
list
in Lisp, *see* Lisp, list
in Prolog, *see* Prolog, list
liveness (of Petri nets), *see* Petri nets, properties, liveness
logbook (in G2), *see* G2, logbook
logical operations, 15
algebraic properties, 16
canonical form, 17
conjunctive normal form, *see* conjunctive normal form
disjunctive normal form, *see* disjunctive normal form
operation table, *see* operation table, of logical operations
logical variable, 129
LOM (last of maxima), *see* fuzzy system, defuzzification methods, LOM
low level Petri nets, *see* Petri nets, low level Petri nets
LTI system, *see* system, LTI system
LTI system models
asymptotic stability, 257
BIBO stability, 257
controllability (state controllability), 255
observability (state observability), 256
stability, 257

matching, 33, 36, 38, 39
in Prolog, *see* Prolog, unification
matrix representation (of Petri nets), *see* Petri nets, matrix representation
max-min composition, *see* fuzzy system, max-min composition
measured data file, 258, 269
measurement device handling, 272
membership function, *see* fuzzy system, membership function
message, 113, 115, 117
in G2, *see* G2, message
message board (in G2), *see* G2, message board
MIMO system, *see* system, MIMO system
modus ponens, 34
generalized, 214
MOM (mean of maxima), *see* fuzzy system, defuzzification methods, MOM

nil, 14
in Lisp , *see* Lisp, list, empty
nonlinear system, *see* system, nonlinear system
NP-complete, *see* algorithmic complexity, NP-complete
NP-hard, *see* algorithmic complexity, NP-complete
number
in Lisp, *see* Lisp, number
in Prolog, *see* Prolog, number

object, 12, 22–25
in G2, *see* G2, object
observability, *see* dynamical properties, observability of LTI systems
open list, 53, 54, 56
operation
interval, 130
logical, *see* logical operations
on fuzzy set, *see* fuzzy system, fuzzy set operation
sign, 129, 145
operation table, 82, 99
interval operations, 131
of confluences, *see* qualitative models, confluences, operation table of
of logical operations, 16
and, 16
implication, 16
or, 17
of sign operations, 129
ordinary Petri nets, *see* Petri nets, ordinary Petri nets
outer product, *see* fuzzy system, outer product

parallelism, *see* Petri nets, parallelism
parameter (in G2), *see* G2, parameter
pattern matching (in Prolog), *see* Prolog, unification
Petri nets, 153–189
analysis, 153, 178–189
invariant analysis, 186–189
reachability tree, 181–186
arc, 154, 162
inhibitor arc, 172–174
k-weighted, 161

boundedness, *see* Petri nets, properties, boundedness
capacity (of place), *see* Petri nets, place, capacity
capacity function, *see* Petri nets, function, capacity function
concurrency, *see* Petri nets, parallelism, concurrency
conflict, *see* Petri nets, parallelism, conflict
confusion, *see* Petri nets, parallelism, confusion
conservation, *see* Petri nets, properties, conservation
conservative Petri nets, 179
coverability, *see* Petri nets, properties, coverability
dead-lock, 180, 185
decomposition, 175
finite capacity Petri nets, 167
firing (of transition), *see* Petri nets, transition, firing
formal definition, 162
function, 154
  capacity function, 167
  marking function, 161, 163, 177
  weight function, 162, 173
graph representation, 153, 154
  arc, *see* Petri nets, arc
  place, *see* Petri nets, place
  transition, *see* Petri nets, transition
incidence matrix, *see* Petri nets, matrix representation
infinite capacity Petri nets, 167
inhibitor arc, *see* Petri nets, arc, inhibitor arc
invariance analysis, *see* Petri nets, analysis, invariance analysis
liveness, *see* Petri nets, properties, liveness
low level Petri nets, 155
marking function, 162, *see* Petri nets, function, marking function
matrix representation, 186
non-primitive transition, *see* Petri nets, transition, non-primitive
ordinary Petri nets, 162
parallelism, 168–172
  concurrency, 168, 171, 186
  conflict, 168, 169, 174, 186
  confusion, 170
place, 154, 162
  capacity, 166
  input, 155
  invariant, *see* Petri nets, properties, place invariant
  output, 155
place invariant, *see* Petri nets, properties, place invariant
primitive transition, *see* Petri nets, transition, primitive
properties, 179–181
  boundedness, 179, 185
  conservation, 179
  coverability, 180, 186
  liveness, 180, 186
  place invariant, 180, 188
  reachability, 180, 186
  safeness, 179, 185
  transition invariant, 180, 188
pure Petri nets, 166, 188
reachability, *see* Petri nets, properties, reachability
reachability set, 177
reachability tree, *see* Petri nets, analysis, reachability tree
safe Petri nets, 179, 185
safeness, *see* Petri nets, properties, safeness
self-loop, 165, 166, 188
sink transition, *see* Petri nets, transition, sink
source transition, *see* Petri nets, transition, source
state-space, 177
token, 155
transition, 154, 162
  dead, 180, 185
  enabled, 158, 162, 167, 173, 177, 185
  firing, 158, 162–165, 172, 176, 177
  invariant, *see* Petri nets, properties, transition invariant
  non-primitive, 176
  primitive, 176
  sink, 165, 166
  source, 165, 166
transition invariant, *see* Petri nets, properties, transition invariant
weight function, *see* Petri nets, function, weight function
place, *see* Petri nets, place
place invariant, *see* Petri nets, properties, place invariant
postprocessing, *see* fuzzy system, fuzzy controller, postprocessing
precondition, *see* condition
predicate, 18, 32
  in Lisp, *see* Lisp, primitive, predicate
  in Prolog, *see* Prolog, predicate

premise, *see* condition
preprocessing, *see* fuzzy system, fuzzy controller, preprocessing
primary data processing, 258
  averaging, 259
  filtering, 259
  handling missing or invalid data, 259
  limit checking, 259
  scaling, 259
primary processing, 272
primary processing data file, 259, 270
primitive, *see* Lisp, primitive
priority, 37, 41, 48, 112, 115, 236
problem reduction, 44
procedure
  in G2, *see* G2, procedure
  in Lisp, *see* Lisp, procedure
  in Prolog, *see* Prolog, procedure
process control functions, 260
  control and regulation, 261
  diagnosis, 261
  identification, 261
  state filtering, 261
process instrumentation diagram, 279
process monitoring functions, 260
  alarm generation, 260
  computation of process trends, 260
  logsheet generation, 260
program
  in Lisp, *see* Lisp, program
  in Prolog, *see* Prolog, program
programming language
  Lisp, *see* Lisp
  Prolog, *see* Prolog
Prolog, 84–103
  atom, 86
  backtracking, 91, 93, 95, 99
  built-in predicate, 96–99
    !, *see* Prolog, built-in predicates, cut
    APPEND, 98
    ASSERTA, 97
    ASSERTZ, 97
    ASSERT, 98
    CONCAT, 98
    FAIL, 99
    NL, 97
    READ, 97
    RETRACT, 97
    WRITE, 97
    cut, 99
    arithmetic, 98
    control, 99
    database handling, 97–98
    expression-handling, 98–99
    input-output, 97
  clause, 85, 87, 94, 96
  database, 85, 97
  fact, 85–86, 88, 91
  goal, 87–90, 95
  Horn clause, *see* Horn clause
  inference engine, 94
  list, 86, 89–90
    empty, 90
    head, 90
    tail, 90
  number, 86
  pattern matching, *see* Prolog, unification
  predicate, 85, 88
    built-in, *see* Prolog, built-in predicate
  procedure, 87
  program, 85, 88, 89
  question, *see* Prolog, goal
  reasoning, 95
  recursion, 96, 100
  relation, 85
  rule, 85, 87–88, 91
    body, 87, 91, 95
    head, 87, 91, 95
  structure, 86, 92
  term, 86, 92
  unification, 90–93
  variable, 86, 92
pure Petri nets, *see* Petri nets, pure Petri nets

QDE, *see* qualitative models, qualitative differential equation (QDE)
QSIM algorithm, *see* qualitative simulation, algorithm
qualitative behaviour, *see* qualitative simulation, qualitative behaviour
qualitative differential equation (QDE), *see* qualitative models, qualitative differential equation (QDE)
qualitative direction, *see* qualitative variable, interval valued, qualitative direction
qualitative equation, 138
qualitative function, 134
  corresponding values, 134
  envelope function, 134
qualitative magnitude, *see* qualitative variable, interval valued, qualitative magnitude
qualitative models, 128, 132–152
  confluences, 145–148
    application of, 147
    operation table of, 145–147
    solution of, 145
  qualitative differential equation (QDE), 137

qualitative differential equation (QDE), 132, 136
Signed Directed Graph (SDG) models, *see* Signed Directed Graph (SDG) models
qualitative physics, 127, 145–148
qualitative reasoning, *see* qualitative reasoning methods
qualitative reasoning methods, 127–152
qualitative physics, *see* qualitative physics
qualitative simulation, *see* qualitative simulation
Signed Directed Graph (SDG) models, *see* Signed Directed Graph (SDG) models
qualitative simulation, 127, 132–144
algorithm, 138, 139
behaviour tree, 142, 144
I-transition, 140–143
landmark set, 133
P-transition, 140–143
qualitative state, 136, 137, 142, 144
qualitative time, *see* qualitative variable, interval valued, qualitative time
qualitative variable, 137, 138, 145, 146
interval valued, 128
landmark set, 137, 138
qualitative direction, 133, 139, 140, 143
qualitative magnitude, 133, 139, 140, 143
qualitative time, 135
logical, *see* logical variable
sign-valued, 128
question (in Prolog), *see* Prolog, goal

raw measured data file, 258, 269
reachability (of Petri nets), *see* Petri nets, properties, reachability
reachability tree, *see* Petri nets, analysis, reachability tree
real-time expert system, *see* expert system, real-time expert system
real-time task, 3
reasoning, 4, 31–51, 85, 104, 113, 191
algorithm, 33
backward, 34, 44–51, 95
bidirectional, 51
counterfactual, 106
explanative, 38, 106
forward, 34, 38–43, 51, 61, 64, 65
hypothetical, 38, 106
in G2, *see* G2, reasoning
in Prolog, *see* Prolog, reasoning
on fuzzy sets, *see* fuzzy system, fuzzy set, inference
qualitative, *see* qualitative reasoning methods
record, 12, 14, 85
recursion
in Lisp, *see* Lisp, recursion
in Prolog, *see* Prolog, recursion
redundancy, *see* fuzzy system, rule-base analysis, redundancy
relation, 14, 18
in G2, *see* G2, relation
in Prolog, *see* Prolog, relation
of fuzzy sets, *see* fuzzy system, fuzzy set, relation
relational database, *see* database, relational
relationship, 12
restriction, *see* fuzzy system, linguistic modifier, restriction
rule, 5, 11, 15, 18–22, 32, 38, 51, 148, 192, 208
condition, 19, 32, 38, 214, 223
consequence, 19, 32, 41
datalog, 19–22, 32, 34, 59, 64
execution, 19, 33
firing, 19, 32, 33, 41
generic (in G2), *see* G2, rule, generic
in fuzzy system, *see* fuzzy system, rule, *see* fuzzy system, rule
in G2, *see* G2, rule
in Prolog, *see* Prolog, rule
semantics, 19
syntax, 19
rule-base, 33, 38, 39, 45, 46, 103
of fuzzy system, *see* fuzzy system, rule-base
normalized, *see* fuzzy system, rule-base, normalized
standard, *see* fuzzy system, rule-base, standard
rule-based system, *see* system, rule-based system

s-curve, *see* fuzzy system, membership function, s-curve
safe Petri nets, *see* Petri nets, safe Petri nets
safeness (of Petri nets), *see* Petri nets, properties, safeness
SDG models, *see* Signed Directed Graph (SDG) models
search, 51–57, 186
A*, 56
blind, 52–54
breadth-first, 54, 57
depth-first, 53, 95
general algorithm, 52
heuristic, 52, 56

hill climbing, 55
informed, *see* search, heuristic
modifiable, 52
non-modifiable, 52, 55
uninformed, *see* search, blind
search graph, 36, 38, 39, 46, 53, 99
search tree, *see* search graph
secondary data processing, *see* primary data processing
secondary processing, 272
semantic nets, 12, 27–29
shouldered curve, *see* fuzzy system, membership function, shouldered curve
sign algebra, 129
sign calculus, 128
sign operation, *see* operation, sign, *see* operation, sign
sign-valued variable, *see* qualitative variable, sign-valued
Signed Directed Graph (SDG) models, 127, 148–152
application of, 151
simulation formula (in G2), *see* G2, simulation formula
simulator (in G2), *see* G2, simulator
singleton, *see* fuzzy system, singleton
SISO system, *see* system, SISO system
stability, *see* dynamical properties, stability of LTI systems
asymptotic stability, *see* dynamical properties, asymptotic stability of LTI systems
BIBO stability, *see* dynamical properties, BIBO stability of LTI systems
state-space, 32, 34, 39, 51
of Petri nets, *see* Petri nets, state-space
state-space model, 127, 132, 253–254
linear(ized), 128, 148
nonlinear, 133, 145, 254
of LTI systems, 148, 254
strong components, *see* dependence graph, strong components
structural controllability, *see* structural dynamical properties, structural controllability
structural dynamical properties
structural controllability, 151
structural observability, 151
structural stability, 151
structural observability, *see* structural dynamical properties, structural observability
structural properties, *see* structural dynamical properties
structural stability, *see* structural dynamical properties, structural stability
structure
in Prolog, *see* Prolog, structure
support (of fuzzy set), *see* fuzzy system, fuzzy set, support
symbol (in Lisp), *see* Lisp, symbol
system
casual system, 253
concentrated parameter system, 254
continuous time system, 253
discrete time system, 253
hierarchical system, 175
linear system, 253
LTI system, 254
MIMO system, 217, 253, 254
nonlinear system, 254
rule-based system, 11
SISO system, 253
time invariant system, 253

tail (of a list)
in Lisp, *see* Lisp, list, tail
in Prolog, *see* Prolog, list, tail
term
in Prolog, *see* Prolog, term
text editor (in G2), *see* G2, text editor
theorem proving (in Prolog), 90
time invariant system, *see* system, time invariant system
token, *see* Petri nets, token
transition, *see* Petri nets, transition
transition invariant, *see* Petri nets, properties, transition invariant
triangular shape curve, *see* fuzzy system, membership function, triangular shape curve
truth table, *see* operation table

unification (in Prolog), *see* Prolog, unification
union (of fuzzy sets), *see* fuzzy system, fuzzy set operation, union
user interface, 6, 31
in expert system, *see* expert system, user interface
in G2, *see* G2, end-user interface

validation, *see* analysis, of rule-base
variable
in G2, *see* G2, variable
in Lisp, *see* Lisp, variable
in Prolog, *see* Prolog, variable
interval-valued, *see* qualitative variable, interval valued
liguistic, *see* fuzzy system, linguistic variable

logical, *see* logical variable
qualitative, *see* qualitative variable
sign-valued, *see* qualitative variable, sign-valued, 129
verification, *see* analysis, of rule-base

workspace (in G2), *see* G2, workspace

z-curve, *see* fuzzy system, membership function, z-curve

# About the Authors

**Katalin Mária Hangos** is currently a Research Professor at the Systems and Control Laboratory of the Computer and Automation Research Institute of the Hungarian Academy of Sciences and a Professor at the Department of Computer Science at University of Veszprém, Hungary. She has been teaching various systems and control related subjects including intelligent control systems, computer controlled systems, system identification and process modelling for more that 5 years for information engineers. Her main interest is dynamic process modelling for control and diagnosis purposes. She is co-author of more that 100 papers on various aspects of modelling and its control applications including nonlinear and stochastic system models, Petri nets, qualitative and graph-theoretic models.

**Rozália Lakner** is currently an Assistant Professor at the Department of Computer Science of University of Veszprém, Hungary. She has been teaching various artificial intelligence related subjects including artificial intelligence, intelligent control systems and process modelling for information engineers. Her main interest is computer-aided dynamic process modelling applying artificial intelligence and computer science methods.

**Miklós Gerzson** is an Associate Professor at the Department of Automation at University of Veszprém. His research interest include modeling and control of different systems, with emphasis on process systems and paralell computing. His teaching activity is related to these fields and to measurement techniques both at University of Veszprém and at University of Pécs. He has authored publications in journals, conference proceedings and undergraduate textbooks.

# Applied Optimization

1. D.-Z. Du and D.F. Hsu (eds.): *Combinatorial Network Theory.* 1996
ISBN 0-7923-3777-8
2. M.J. Panik: *Linear Programming: Mathematics, Theory and Algorithms.* 1996
ISBN 0-7923-3782-4
3. R.B. Kearfott and V. Kreinovich (eds.): *Applications of Interval Computations.* 1996 ISBN 0-7923-3847-2
4. N. Hritonenko and Y. Yatsenko: *Modeling and Optimimization of the Lifetime of Technology.* 1996 ISBN 0-7923-4014-0
5. T. Terlaky (ed.): *Interior Point Methods of Mathematical Programming.* 1996
ISBN 0-7923-4201-1
6. B. Jansen: *Interior Point Techniques in Optimization.* Complementarity, Sensitivity and Algorithms. 1997 ISBN 0-7923-4430-8
7. A. Migdalas, P.M. Pardalos and S. Storøy (eds.): *Parallel Computing in Optimization.* 1997 ISBN 0-7923-4583-5
8. F.A. Lootsma: *Fuzzy Logic for Planning and Decision Making.* 1997
ISBN 0-7923-4681-5
9. J.A. dos Santos Gromicho: *Quasiconvex Optimization and Location Theory.* 1998
ISBN 0-7923-4694-7
10. V. Kreinovich, A. Lakeyev, J. Rohn and P. Kahl: *Computational Complexity and Feasibility of Data Processing and Interval Computations.* 1998
ISBN 0-7923-4865-6
11. J. Gil-Aluja: *The Interactive Management of Human Resources in Uncertainty.* 1998
ISBN 0-7923-4886-9
12. C. Zopounidis and A.I. Dimitras: *Multicriteria Decision Aid Methods for the Prediction of Business Failure.* 1998 ISBN 0-7923-4900-8
13. F. Giannessi, S. Komlósi and T. Rapcsák (eds.): *New Trends in Mathematical Programming.* Homage to Steven Vajda. 1998 ISBN 0-7923-5036-7
14. Ya-xiang Yuan (ed.): *Advances in Nonlinear Programming.* Proceedings of the '96 International Conference on Nonlinear Programming. 1998 ISBN 0-7923-5053-7
15. W.W. Hager and P.M. Pardalos: *Optimal Control.* Theory, Algorithms, and Applications. 1998 ISBN 0-7923-5067-7
16. Gang Yu (ed.): *Industrial Applications of Combinatorial Optimization.* 1998
ISBN 0-7923-5073-1
17. D. Braha and O. Maimon (eds.): *A Mathematical Theory of Design: Foundations, Algorithms and Applications.* 1998 ISBN 0-7923-5079-0

# Applied Optimization

18. O. Maimon, E. Khmelnitsky and K. Kogan: *Optimal Flow Control in Manufacturing.* Production Planning and Scheduling. 1998 ISBN 0-7923-5106-1
19. C. Zopounidis and P.M. Pardalos (eds.): *Managing in Uncertainty: Theory and Practice.* 1998 ISBN 0-7923-5110-X
20. A.S. Belenky: *Operations Research in Transportation Systems:* Ideas and Schemes of Optimization Methods for Strategic Planning and Operations Management. 1998 ISBN 0-7923-5157-6
21. J. Gil-Aluja: *Investment in Uncertainty.* 1999 ISBN 0-7923-5296-3
22. M. Fukushima and L. Qi (eds.): *Reformulation: Nonsmooth, Piecewise Smooth, Semismooth and Smooting Methods.* 1999 ISBN 0-7923-5320-X
23. M. Patriksson: *Nonlinear Programming and Variational Inequality Problems.* A Unified Approach. 1999 ISBN 0-7923-5455-9
24. R. De Leone, A. Murli, P.M. Pardalos and G. Toraldo (eds.): *High Performance Algorithms and Software in Nonlinear Optimization.* 1999 ISBN 0-7923-5483-4
25. A. Schöbel: *Locating Lines and Hyperplanes.* Theory and Algorithms. 1999 ISBN 0-7923-5559-8
26. R.B. Statnikov: *Multicriteria Design.* Optimization and Identification. 1999 ISBN 0-7923-5560-1
27. V. Tsurkov and A. Mironov: *Minimax under Transportation Constrains.* 1999 ISBN 0-7923-5609-8
28. V.I. Ivanov: *Model Development and Optimization.* 1999 ISBN 0-7923-5610-1
29. F.A. Lootsma: *Multi-Criteria Decision Analysis via Ratio and Difference Judgement.* 1999 ISBN 0-7923-5669-1
30. A. Eberhard, R. Hill, D. Ralph and B.M. Glover (eds.): *Progress in Optimization.* Contributions from Australasia. 1999 ISBN 0-7923-5733-7
31. T. Hürlimann: *Mathematical Modeling and Optimization.* An Essay for the Design of Computer-Based Modeling Tools. 1999 ISBN 0-7923-5927-5
32. J. Gil-Aluja: *Elements for a Theory of Decision in Uncertainty.* 1999 ISBN 0-7923-5987-9
33. H. Frenk, K. Roos, T. Terlaky and S. Zhang (eds.): *High Performance Optimization.* 1999 ISBN 0-7923-6013-3
34. N. Hritonenko and Y. Yatsenko: *Mathematical Modeling in Economics, Ecology and the Environment.* 1999 ISBN 0-7923-6015-X
35. J. Virant: *Design Considerations of Time in Fuzzy Systems.* 2000 ISBN 0-7923-6100-8

# Applied Optimization

36. G. Di Pillo and F. Giannessi (eds.): *Nonlinear Optimization and Related Topics*. 2000 ISBN 0-7923-6109-1
37. V. Tsurkov: *Hierarchical Optimization and Mathematical Physics*. 2000 ISBN 0-7923-6175-X
38. C. Zopounidis and M. Doumpos: *Intelligent Decision Aiding Systems Based on Multiple Criteria for Financial Engineering*. 2000 ISBN 0-7923-6273-X
39. X. Yang, A.I. Mees, M. Fisher and L.Jennings (eds.): *Progress in Optimization*. Contributions from Australasia. 2000 ISBN 0-7923-6286-1
40. D. Butnariu and A.N. Iusem: *Totally Convex Functions for Fixed Points Computation and Infinite Dimensional Optimization*. 2000 ISBN 0-7923-6287-X
41. J. Mockus: *A Set of Examples of Global and Discrete Optimization*. Applications of Bayesian Heuristic Approach. 2000 ISBN 0-7923-6359-0
42. H. Neunzert and A.H. Siddiqi: *Topics in Industrial Mathematics*. Case Studies and Related Mathematical Methods. 2000 ISBN 0-7923-6417-1
43. K. Kogan and E. Khmelnitsky: *Scheduling: Control-Based Theory and Polynomial-Time Algorithms*. 2000 ISBN 0-7923-6486-4
44. E. Triantaphyllou: *Multi-Criteria Decision Making Methods*. A Comparative Study. 2000 ISBN 0-7923-6607-7
45. S.H. Zanakis, G. Doukidis and C. Zopounidis (eds.): *Decision Making: Recent Developments and Worldwide Applications*. 2000 ISBN 0-7923-6621-2
46. G.E. Stavroulakis: *Inverse and Crack Identification Problems in Engineering Mechanics*. 2000 ISBN 0-7923-6690-5
47. A. Rubinov and B. Glover (eds.): *Optimization and Related Topics*. 2001 ISBN 0-7923-6732-4
48. M. Pursula and J. Niittymäki (eds.): *Mathematical Methods on Optimization in Transportation Systems*. 2000 ISBN 0-7923-6774-X
49. E. Cascetta: *Transportation Systems Engineering: Theory and Methods*. 2001 ISBN 0-7923-6792-8
50. M.C. Ferris, O.L. Mangasarian and J.-S. Pang (eds.): *Complementarity: Applications, Algorithms and Extensions*. 2001 ISBN 0-7923-6816-9
51. V. Tsurkov: *Large-scale Optimization – Problems and Methods*. 2001 ISBN 0-7923-6817-7
52. X. Yang, K.L. Teo and L. Caccetta (eds.): *Optimization Methods and Applications*. 2001 ISBN 0-7923-6866-5
53. S.M. Stefanov: *Separable Programming Theory and Methods*. 2001 ISBN 0-7923-6882-7

# Applied Optimization

54. S.P. Uryasev and P.M. Pardalos (eds.): *Stochastic Optimization: Algorithms and Applications.* 2001 ISBN 0-7923-6951-3
55. J. Gil-Aluja (ed.): *Handbook of Management under Uncertainty.* 2001 ISBN 0-7923-7025-2
56. B.-N. Vo, A. Cantoni and K.L. Teo: *Filter Design with Time Domain Mask Constraints: Theory and Applications.* 2001 ISBN 0-7923-7138-0
57. S. Zlobec: *Stable Parametric Programming.* 2001 ISBN 0-7923-7139-9
58. M.G. Nicholls, S. Clarke and B. Lehaney (eds.): *Mixed-Mode Modelling: Mixing Methodologies for Organisational Intervention.* 2001 ISBN 0-7923-7151-8
59. F. Giannessi, P.M. Pardalos and T. Rapcsák (eds.): *Optimization Theory. Recent Developments from Mátraháza.* 2001 ISBN 1-4020-0009-X
60. K.M. Hangos, R. Lakner and M. Gerzson: *Intelligent Control Systems.* An Introduction with Examples. 2001 ISBN 1-4020-0134-7

KLUWER ACADEMIC PUBLISHERS – DORDRECHT / BOSTON / LONDON